Jossey-Bass Teacher

Jossey-Bass Teacher provides K–12 teachers with essential knowledge and tools to create a positive and lifelong impact on student learning. Trusted and experienced educational mentors offer practical classroom-tested and theory-based teaching resources for improving teaching practice in a broad range of grade levels and subject areas. From one educator to another, we want to be your first source to make every day your best day in teaching. *Jossey-Bass Teacher* resources serve two types of informational needs—essential knowledge and essential tools.

Essential knowledge resources provide the foundation, strategies, and methods from which teachers may design curriculum and instruction to challenge and excite their students. Connecting theory to practice, essential knowledge books rely on a solid research base and time-tested methods, offering the best ideas and guidance from many of the most experienced and well-respected experts in the field.

Essential tools save teachers time and effort by offering proven, ready-to-use materials for in-class use. Our publications include activities, assessments, exercises, instruments, games, ready reference, and more. They enhance an entire course of study, a weekly lesson, or a daily plan. These essential tools provide insightful, practical, and comprehensive materials on topics that matter most to K–12 teachers.

The Math Teacher's
BOOK OF LISTS

SECOND EDITION

JUDITH A. MUSCHLA

GARY ROBERT MUSCHLA

JOSSEY-BASS
A Wiley Imprint
www.josseybass.com

Published by Jossey-Bass
A Wiley Imprint
989 Market Street, San Francisco, CA 94103-1741   www.josseybass.com

Jossey-Bass books and products are available through most bookstores. To contact Jossey-Bass directly call our Customer Care Department within the U.S. at 800-956-7739, outside the U.S. at 317-572-3986, or fax 317-572-4002.

Jossey-Bass also publishes its books in a variety of electronic formats. Some content that appears in print may not be available in electronic books.

The fact that an organization or Web site is referred to in this work as a citation and/or a potential source of further information does not mean that the author or the publisher endorses the information the organization or Web site may provide or recommendations it may make. Further, readers should be aware that Internet Web sites listed in this work may have changed or disappeared between when this work was written and when it is read.

Library of Congress Cataloging-in-Publication Data
Muschla, Judith A.
  The math teacher's book of lists / Judith A. Muschla and Gary Robert Muschla.— 2nd ed.
    p. cm.
  ISBN 0-7879-7398-X (alk. paper)
  1. Mathematics—Study and teaching. I. Muschla, Gary Robert. II. Title.
  QA11.2.M88 2005
  510'.71'2—dc22
                              2004024470
Printed in the United States of America
SECOND EDITION
PB Printing 10 9 8 7 6 5 4 3 2

About This Book

Appropriate for grade levels 5 through 12, *The Math Teacher's Book of Lists* (2nd ed.) consists of ten sections. Sections One through Nine contain reproducible lists and offer specific information on over 300 topics. Section Ten contains a variety of reproducible teaching aids that can be used as necessary in support of a teacher's instructional program.

This new edition retains all of the strengths of the first edition while expanding its scope and relevance. The original lists have been updated, obsolete lists have been dropped, and several new lists, including "Fractal Facts," "Topics in Discrete Math," "Math Web Sites for Students," and "Math Web Sites for Teachers," have been added. A new section, "Special Reference Lists for Students," is also included.

The new edition, like the original, is designed for easy implementation. Each list is written in clear, simple-to-read language, stands alone, and may be used with students of various grades and abilities, enabling teachers to customize the materials to their needs.

The lists can be used to introduce or review information, develop supplementary materials for the classroom, or expand topics in the curriculum. Moreover, the lists are cross-referenced, allowing teachers to extend lessons with related topics.

The lists throughout this book will help teachers to show their students the broad range of mathematics, recognize math's relevance to today's world, and support students in mastering the skills essential to their courses of study.

About the Authors

Gary Robert Muschla received his B.A. and M.A.T. from Trenton State College and taught at Appleby School in Spotswood, New Jersey, for more than twenty-five years. He spent many of his years in the classroom teaching mathematics at the middle school level. He has also taught reading and writing and has been a successful author. He is a member of the Authors Guild.

Mr. Muschla has written several resources for teachers, including: *English Teacher's Great Books Activities Kit* (1994), *Reading Workshop Survival Kit* (1997), *Ready-to-Use Reading Proficiency Lessons and Activities, Fourth-Grade Level* (2002), *Ready-to-Use Reading Proficiency Lessons and Activities, Eighth-Grade Level* (2002), *Ready-to-Use Reading Proficiency Lessons and Activities, Tenth-Grade Level* (2003), *The Writing Teacher's Book of Lists* (2nd ed., 2004), and *Writing Workshop Survival Kit* (2nd ed., 2005), all published by Jossey-Bass. He currently writes and serves as a consultant in education.

Judith A. Muschla received her B.A. in mathematics from Douglass College at Rutgers University and is certified to teach K–12. She has taught mathematics in South River, New Jersey, for over twenty-five years. She has taught math at various levels at South River High School, ranging from basic skills through precalculus. She has also taught at South River Middle School, where, in her capacity as a team leader, she helped revise the mathematics curriculum to reflect the Standards of the National Council of Teachers of Mathematics, coordinated interdisciplinary units, and conducted mathematics workshops for teachers and parents. She was a recipient of the 1990–1991 Governor's Teacher Recognition Program Award in New Jersey, and she was named the 2002 South River Public School District Teacher of the Year. Along with teaching and writing, she has been a member of the state Standards Review Panel for the Mathematics Core Curriculum Content Standards in New Jersey.

Along with this revised edition of *The Math Teacher's Book of Lists,* Gary and Judith Muschla have also coauthored *Hands-On Math Projects with Real-Life Applications* (1996), *Math Starters! 5- to 10-Minute Activities to Make Kids Think, Grades 6–12* (1999), *The Geometry Teacher's Activities Kit* (2000), *Math Smart! Over 220 Ready-to-Use Activities to Motivate and Challenge Students, Grades 6–12* (2002), *The Algebra Teacher's Activities Kit: 150 Ready-to-Use Activities with Real-World Applications* (2003), and *Math Games: 180 Reproducible Activities to Motivate, Excite, and Challenge Students, Grades 6–12* (2004), all published by Jossey-Bass.

For Erin

Acknowledgments

We thank Paul Coleman, principal of South River High School in South River, New Jersey, and our colleagues for their support of our efforts in writing this book.

We also thank Steve Thompson, our editor, for his encouragement and support from the beginning of this project to its completion.

Thanks also to our daughter, Erin, who was the first reader of the manuscript and who caught our initial typos, oversights, and omissions.

We acknowledge the efforts of the late Richard Kwiatkowski, former Math Department chairman at South River High, for his suggestions on our original manuscript. Thanks also to Sonia Helton, professor of education at the University of South Florida, whose comments on the original manuscript were both helpful and informative.

We appreciate the efforts of Michele Quiroga, our production editor, for her efforts in guiding the manuscript through the production process. Special thanks to Diane Turso for proofreading our work and catching all those little things we didn't catch.

Finally, we thank our students, who over the years have made teaching challenging, exciting, and rewarding.

Preface

Since the publication of the first edition of this book in 1995, the importance of mathematics has never been greater. The world today runs on numbers. Make a budget, buy a car, choose the best cell phone plan, design a building, or determine the best flight path for a probe to Mars, and you are using mathematics. Math is everywhere, and its importance will only continue to grow.

To ensure that the mathematics taught in their schools is relevant to the needs of their students, math teachers across the country regularly revise and update their curriculums. Increasingly, the goal of mathematics instruction is to help students apply math to solve real-life problems and understand their world. In order to implement the Standards of the National Council of Teachers of Mathematics, math teachers should conduct classes in which students investigate meaningful problems, work in groups and share ideas and insights, examine models, use calculators and computers in problem solving, write about their observations and conclusions, and connect math with other subject areas.

As a mathematics teacher, you work side by side with your students, encouraging and supporting their efforts in mastering math, and helping them to acquire the skills that will serve them well in the years to come. To those goals, we trust that this book will be an essential resource.

How to Use This Book

This second edition of *The Math Teacher's Book of Lists* contains nine sections of 306 reproducible lists. Section Ten offers an assortment of handy reproducible teaching aids that you can use as needed in your program. All of the lists and teaching aids can be adapted to various methods of instruction, giving you great flexibility. Following is a brief summary of each section.

Section One, "Numbers: Theory and Operation," contains 40 lists. Among the topics covered are prime and composite numbers; perfect squares and cubes; rules for divisibility; rules for finding the greatest common factor and least common multiple; rules for working with fractions, decimals, percents, and integers; rules for rounding; scientific notation; and mathematical signs and symbols.

Section Two, "Measurement," contains 35 lists. The topics include obsolete units of measure; units of the English system and metric system; English–metric equivalents; conversion factors for length, area, and volume; liquid capacity; dry capacity and mass; time; temperature; wind speed; and the Richter Scale.

Section Three, "Geometry," contains 58 lists. The lists in this section focus on lines, planes, angles, polygons, triangles, proportions, quadrilaterals, tessellations, circles, arcs, solids, symmetry, fractals, area, and volume.

Section Four, "Algebra," contains 49 lists. These lists cover properties of real numbers; axioms of order and equality; rules for exponents; steps for solving equations; polynomials; common factoring formulas; linear equations; types of functions; square roots; quadratic equations; graphs; conic sections; complex numbers; vectors; and matrices.

Section Five, "Trigonometry and Calculus," contains 37 lists. Topics focus on trigonometric functions; basic trigonometric functions and their graphs; vertical and horizontal line tests; limits; continuity; differentiation rules; high-order derivatives; integration rules; definite integrals; and natural logarithmic functions.

Section Six, "Math in Other Areas," contains 33 lists. Topics include types of graphs, displays, and tables; topics in discrete math; permutations; combinations; probability; odds; writing and endorsing a check; loans and interest; steps to making a budget; stock market words; reading a financial page; the numbers in popular sports; mathematical facts about space; conservation facts; and some helpful formulas.

Section Seven, "Potpourri," contains 16 lists. The lists of this section include the history of mathematics; famous mathematicians through history; careers in math; mathematical palindromes; numbers and symbolism; math and superstition; mathematical idioms; numbers and language; and the Greek alphabet.

Section Eight, "Special Reference Lists for Students," contains 10 lists. Topics here include the math student's responsibilities; overcoming math anxiety; strategies for taking math tests; problem-solving strategies; math journal guidelines for students; and math Web sites for students.

Section Nine, "Lists for Teachers' Reference," contains 28 lists. Topics include the math teacher's responsibilities; upgrading your mathematics curriculum; management strategies; managing a cooperative math class; steps for conducting effective conferences with students and with parents; sources of problems-of-the-day; portfolios; helping students prepare for math tests; how to produce a math magazine; possible guest speakers for your math class; steps for maintaining positive relations with parents; how to have a parents' math night; math Web sites for teachers; and a self-appraisal for math teachers.

Because each list contains specific information about a particular topic or skill, the lists may be used to introduce lessons, reinforce concepts, synthesize related topics, or as reviews before tests or quizzes. The lists in Sections Six and Seven provide a wealth of information that you or your students can use to create word problems or use as databases for math projects. If you wish, you might instruct students to file the lists you distribute to them in folders according to general topic for future reference. Not only will they be able to refer to the lists as needed, but the saved lists will build an impressive reference collection over the year.

In **Section Ten, "Reproducible Teaching Aids,"** you will find such classroom aids as number lines; grids of various dimensions; isometric dot paper; decimal squares; fraction strips; algebra tiles; tangrams; grids for trig functions; and a variety of nets.

We suggest you use this book as a resource, choosing the lists you need to supplement your curriculum and teaching methods. The material will help you to enhance your math program and make your teaching easier and more effective.

Judy and Gary Muschla

Contents

Section One: Numbers: Theory and Operation

Section Two: Measurement

Section Three: Geometry

Section Four: Algebra

Section Five: Trigonometry and Calculus

Section Six: Math in Other Areas

Section Seven: Potpourri

Section Eight: Special Reference Lists for Students

Section Nine: Lists for Teachers' Reference

Section Ten: Reproducible Teaching Aids

Numbers:
Theory and Operation

The Real Numbers

Numbers and the operations we can perform with them are the basis of our numerical system. Each set of numbers listed below is a part or subset of the *Real Numbers,* which is also known as the *Real Number System* or the *Continuum of Real Numbers.*

▶ *Natural Numbers*—the set of the counting numbers. They can be classified as odd or even. {1, 2, 3, 4, 5 . . .}

▶ *Whole Numbers*—the set of the natural numbers and zero. {0, 1, 2, 3, 4, 5 . . .}

▶ *Integers*—the set of the natural numbers, their opposites, and zero. {. . . –3, –2, –1, 0, 1, 2, 3 . . .}

▶ *Rational Numbers*—the set of all the numbers that can be expressed in the form *a/b* where *a* and *b* are integers, $b \neq 0$. *Examples:* integers, finite decimals and their opposites, and repeating decimals and their opposites.

▶ *Irrational Numbers*—the set of the numbers that cannot be written as terminating or repeating decimals. *Examples:* $\sqrt{2}$, $\sqrt{3}$, π, and *e*.

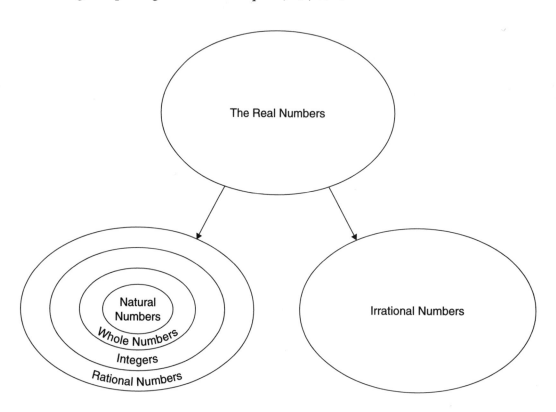

LIST 2

Classification of Real Numbers

In life, it is common to classify things that have common characteristics. In math, we can classify real numbers as shown by the following chart.

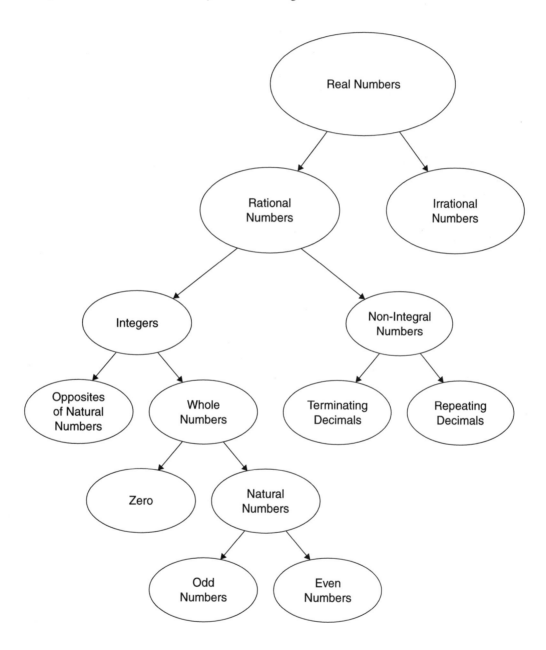

LIST 3

Cardinal and Ordinal Numbers

When we count, or specify a total number of items, we are using *cardinal numbers*. For example, *ten* is a cardinal number. *Ordinal numbers* are used to show order. *Tenth* is an ordinal number. Several examples of cardinal and ordinal numbers appear below. The shortened forms of ordinal numbers are also shown.

Cardinal Number	Ordinal Number	Shortened Form
1	First	1st
2	Second	2nd
3	Third	3rd
4	Fourth	4th
5	Fifth	5th
6	Sixth	6th
7	Seventh	7th
8	Eighth	8th
9	Ninth	9th
10	Tenth	10th
11	Eleventh	11th
12	Twelfth	12th
13	Thirteenth	13th
14	Fourteenth	14th
15	Fifteenth	15th
16	Sixteenth	16th
17	Seventeenth	17th
18	Eighteenth	18th
19	Nineteenth	19th
20	Twentieth	20th
21	Twenty-first	21st
22	Twenty-second	22nd
23	Twenty-third	23rd
24	Twenty-fourth	24th
25	Twenty-fifth	25th
26	Twenty-sixth	26th
27	Twenty-seventh	27th
28	Twenty-eighth	28th
29	Twenty-ninth	29th
30	Thirtieth	30th

Successive numbers repeat in the same manner:

100	One hundredth	100th
101	One hundred first	101st
102	One hundred second	102nd
103	One hundred third	103rd
104	One hundred fourth	104th
105	One hundred fifth	105th

LIST 4

Prime Numbers

A *prime number* is an integer greater than 1 whose only whole number factors are itself and 1. For example, 2 is a prime number because its only factors are 2 and 1. The number 383 is prime for the same reason. Its only factors are 383 and 1. Following is a list of the prime numbers less than 1,000:

2	127	283	467	661	877
3	131	293	479	673	881
5	137	307	487	677	883
7	139	311	491	683	887
11	149	313	499	691	907
13	151	317	503	701	911
17	157	331	509	709	919
19	163	337	521	719	929
23	167	347	523	727	937
29	173	349	541	733	941
31	179	353	547	739	947
37	181	359	557	743	953
41	191	367	563	751	967
43	193	373	569	757	971
47	197	379	571	761	977
53	199	383	577	769	983
59	211	389	587	773	991
61	223	397	593	787	997
67	227	401	599	797	
71	229	409	601	809	
73	233	419	607	811	
79	239	421	613	821	
83	241	431	617	823	
89	251	433	619	827	
97	257	439	631	829	
101	263	443	641	839	
103	269	449	643	853	
107	271	457	647	857	
109	277	461	653	859	
113	281	463	659	863	

Types of Prime Numbers

Prime numbers are *natural numbers* that have only two factors: 1 and the number. Certain types of primes have common characteristics. These primes are grouped together and have special names.

Twin primes are a set of two consecutive odd primes, which differ by only two. The twin primes less than 100 follow:

3,5	29,31
5,7	41,43
11,13	59,61
17,19	71,73

Symmetric primes, also called *Euler primes,* are a pair of prime numbers that are the same distance from a given number on a number line. There are no symmetric primes for 1, 2, or 3. It has not been proven if all natural numbers greater than 3 have symmetric primes. The following list shows the symmetric primes for the numbers 1 through 25.

Number	Symmetric Primes
1	None
2	None
3	None
4	3,5
5	3,7
6	5,7
7	3,11
8	5,11; 3,13
9	7,11; 5,13
10	7,13; 3,17
11	5,17; 3,19
12	11,13; 7,17; 5,19
13	7,19; 3,23
14	11,17; 5,23
15	13,17; 11,19; 7,23
16	15,17; 13,19; 3,29
17	11,23; 5,29; 3,31
18	17,19; 13,23; 7,29; 5,31
19	9,29; 7,31
20	17,23; 11,29; 3,37
21	19,23; 13,29; 11,31; 5,37
22	13,31; 7,37; 3,41
23	17,29; 13,33; 5,41; 3,43
24	19,29; 17,31; 11,37; 7,41; 5,43
25	19,31; 13,37; 7,43; 3,47

LIST 5

(Continued)

An *emirp* is a prime number that remains a prime when its digits are reversed. *Emirp*, of course, is *prime* spelled backward. Following are the emirps less than 500:

11	13	17	31	37	71	73	79	97	101
107	113	131	149	151	157	167	179	181	191
199	311	313	337	347	353	359	373	383	389

Relatively prime numbers are numbers whose greatest common factor is 1. If two numbers are relatively prime, they are said to be relatively prime in pairs. Notice that the numbers of the following list are not limited to primes. Numbers that are relatively prime do not have to be prime numbers. They must not have any common factors other than 1. Below is a list of the numbers from 1 through 10 that are relatively prime in pairs:

1,2	2,3	3,4	4,5	5,6
1,3	2,5	3,5	4,7	5,7
1,4	2,7	3,7	4,9	5,8
1,5	2,9	3,8		5,9
1,6		3,10		
1,7				
1,8				
1,9				
1,10				
6,7	7,8	8,9	9,10	
	7,9			
	7,10			

LIST 6

Composite Numbers

Composite numbers are positive integers that have more than two positive whole number factors. The number 6 is a composite number because its factors are 1, 2, 3, and 6. Note that 1 is the only natural number that is neither prime nor composite. The first 100 composite numbers follow:

4	39	72	104
6	40	74	105
8	42	75	106
9	44	76	108
10	45	77	110
12	46	78	111
14	48	80	112
15	49	81	114
16	50	82	115
18	51	84	116
20	52	85	117
21	54	86	118
22	55	87	119
24	56	88	120
25	57	90	121
26	58	91	122
27	60	92	123
28	62	93	124
30	63	94	125
32	64	95	126
33	65	96	128
34	66	98	129
35	68	99	130
36	69	100	132
38	70	102	133

LIST 7

Perfect Squares and Cubes

The *square* of a number results when a number is multiplied by itself—for example, $4 \times 4 = 16$. Thus we say 4 squared is 16.

The *cube* of a number results when a number is multiplied by itself twice—for example, $4 \times 4 \times 4 = 64$. Thus, we say that 4 cubed is 64.

The following list contains the squares and cubes of numbers 1 to 25:

Number	Squared	Cubed
1	1	1
2	4	8
3	9	27
4	16	64
5	25	125
6	36	216
7	49	343
8	64	512
9	81	729
10	100	1,000
11	121	1,331
12	144	1,728
13	169	2,197
14	196	2,744
15	225	3,375
16	256	4,096
17	289	4,913
18	324	5,832
19	361	6,859
20	400	8,000
21	441	9,261
22	484	10,648
23	529	12,167
24	576	13,824
25	625	15,625

LIST 8

Abundant, Deficient, and Perfect Numbers

The ancient Greeks were thinkers of the first order. They enjoyed mathematics and categorized all the natural numbers as being abundant, deficient, or perfect:

Abundant—a number that is less than the sum of its factors, excluding itself

Deficient—a number that is greater than the sum of its factors, excluding itself

Perfect—a number that is equal to the sum of its factors, excluding itself

The first list that follows groups the first 50 numbers as the ancient Greeks might have. The second list contains perfect numbers.

Number	Factors Excluding Itself	Sum	Type
1		0	Deficient
2	1	1	Deficient
3	1	1	Deficient
4	1,2	3	Deficient
5	1	1	Deficient
6	1,2,3	6	Perfect
7	1	1	Deficient
8	1,2,4	7	Deficient
9	1,3	4	Deficient
10	1,2,5	8	Deficient
11	1	1	Deficient
12	1,2,3,4,6	16	Abundant
13	1	1	Deficient
14	1,2,7	10	Deficient
15	1,3,5	9	Deficient
16	1,2,4,8	15	Deficient
17	1	1	Deficient
18	1,2,3,6,9	21	Abundant
19	1	1	Deficient
20	1,2,4,5,10	22	Abundant
21	1,3,7	11	Deficient
22	1,2,11	14	Deficient
23	1	1	Deficient
24	1,2,3,4,6,8,12	36	Abundant
25	1,5	6	Deficient
26	1,2,13	16	Deficient
27	1,3,9	13	Deficient
28	1,2,4,7,14	28	Perfect
29	1	1	Deficient
30	1,2,3,5,6,10,15	42	Abundant

LIST 8

(Continued)

Number	Factors Excluding Itself	Sum	Type
31	1	1	Deficient
32	1,2,4,8,16	31	Deficient
33	1,3,11	15	Deficient
34	1,2,17	20	Deficient
35	1,5,7	13	Deficient
36	1,2,3,4,6,9,12,18	55	Abundant
37	1	1	Deficient
38	1,2,19	22	Deficient
39	1,3,13	17	Deficient
40	1,2,4,5,8,10,20	50	Abundant
41	1	1	Deficient
42	1,2,3,6,7,14,21	54	Abundant
43	1	1	Deficient
44	1,2,4,11,22	40	Deficient
45	1,3,5,9,15	33	Deficient
46	1,2,23	26	Deficient
47	1	1	Deficient
48	1,2,3,4,6,8,12,16,24	76	Abundant
49	1,7	8	Deficient
50	1,2,5,10,25	43	Deficient

Perfect Numbers

Perfect numbers are mathematical rarities that have no practical use. Still, mathematicians have found them challenging and have even worked out a formula to find them: $2^{p-1}(2^p - 1)$ where p and $(2^p - 1)$ are prime numbers. No one has found an odd perfect number, and no one has proven whether odd perfect numbers exist. A list of the first eight perfect numbers and their formulas follows. After that the numbers simply become too large to write.

Perfect Number	Formula
6	$2^1(2^2 - 1)$
28	$2^2(2^3 - 1)$
496	$2^4(2^5 - 1)$
8,128	$2^6(2^7 - 1)$
33,550,336	$2^{12}(2^{13} - 1)$
8,589,869,056	$2^{16}(2^{17} - 1)$
137,438,691,328	$2^{18}(2^{19} - 1)$
2,305,843,008,139,952,128	$2^{30}(2^{31} - 1)$

LIST **9**

Amicable Numbers

Amicable numbers come in pairs. They are quite special, because each number of the pair has factors that (excluding itself) add up to equal the other number of the pair.

The first pair of amicable numbers is 220 and 284. The factors of 220, excluding itself, are 1, 2, 4, 5, 10, 11, 20, 22, 44, 55, and 110. Their sum equals 284. The factors of 284, excluding itself, are 1, 2, 4, 71, and 142, which add up to 220.

More than 1,000 pairs of amicable numbers have been found. This list offers the ten smallest pairs:

220 and 284

1,184 and 1,210

2,620 and 2,924

5,020 and 5,564

6,232 and 6,368

10,744 and 10,856

12,285 and 14,595

17,296 and 18,416

63,020 and 76,084

66,928 and 66,992

LIST 10

Some Lesser-Known Types of Numbers

Mathematicians often group numbers according to their properties. While most students of math are familiar with the types of numbers found in List 1, "The Real Numbers," there are many others that are not so well known.

Algebraic Numbers—numbers that are the solution of an algebraic equation. *Example:* $2 + x = 9$. The answer is 7, which is an algebraic number.

Almost Perfect Numbers—numbers that are one more or one less than the sum of all of their factors, except themselves. *Example:* 4 is almost perfect because $1 + 2 = 3$, which is one less than 4. (All powers of 2 are almost perfect.)

Automorphic Numbers—numbers that when raised to a power end in the original number. *Example:* $5^2 = 25$; $5^3 = 125$; $6^2 = 36$; $25^2 = 625$.

Complex Numbers—numbers in the form $a + bi$ where a and b stand for real numbers and $i^2 = -1$ or $(i = \sqrt{-1})$. *Example:* $-2 + \sqrt{-2}$ can be written as $-2 + i\sqrt{2}$.

Crowd—a chain of three sociable numbers. No one has yet found such a chain, but neither has anyone proven that such chains do not exist.

Cute Numbers—numbers that have exactly four factors. *Examples:* 6, whose factors are 1, 2, 3, and 6; 8, whose factors are 1, 2, 4, and 8.

Cyclic Numbers—an integer of n digits with the following characteristic: When multiplied by a number from 1 to n, the product has the same digits as the original number, in the same cycle. *Example:* 142,857 is the smallest cyclic number, other than 1.

$$1 \times 142,857 = 142,857$$
$$2 \times 142,857 = 285,714$$
$$3 \times 142,857 = 428,571$$
$$4 \times 142,857 = 571,428$$
$$5 \times 142,857 = 714,285$$
$$6 \times 142,857 = 857,142$$

Denominate Numbers—numbers whose unit represents a unit of measure. *Examples:* 3 pounds, 7 inches, 2 quarts.

Fibonacci Numbers—numbers of the sequence 1, 1, 2, 3, 5, etc., . . . where successive numbers are the sum of the two preceding numbers. *Examples:* $1 + 1 = 2$; $1 + 2 = 3$; $2 + 3 = 5$.

Imaginary Numbers—numbers involving the imaginary unit i, where $i^2 = -1$ or $(i = \sqrt{-1})$. *Example:* $\sqrt{-2} = i\sqrt{2}$.

Lucas Numbers—numbers of the sequence 1, 3, 4, 7, 11, 18, . . . where successive numbers are the sum of the two preceding numbers. *Examples:* $1 + 3 = 4$; $3 + 4 = 7$; $4 + 7 = 11$; $7 + 11 = 18$; etc.

Mersenne Primes—primes of the form $2^p - 1$ where p is prime. *Examples:* 3, 7, 31, 127.

LIST 10

(Continued)

Random Numbers—numbers that are obtained without any pattern and can be described only by listing the digits. *Example:* Picking numbers from a hat, such as 8, 3, 7, 0, 1, 2, 9, 9, 8, 1, 2, and 6.

Repunit Numbers—an abbreviated form of "repeated unit" of the digit 1, but excluding the number 1. *Examples:* 11, 111, 1111, etc.

Sociable Numbers—one of a chain of numbers whose factors add up to the next number in the chain. *Example:* a five-link chain: 12496, 14288, 15472, 14536, 14264.

The sum of the factors of 12496 = 14288.

The sum of the factors of 14288 = 15472.

The sum of the factors of 15472 = 14536.

The sum of the factors of 14536 = 14264.

The sum of the factors of 14264 = 12496.

Note that the chain is completed with the last number.

Surds—algebraic numbers that cannot be written as an exact ratio of two integers. A surd is one type of irrational number. (The other type of irrational number is the transcendental number.) *Examples:* $\sqrt{2}, \sqrt{3}$.

Transcendental Numbers—any irrational numbers that are not algebraic numbers. *Examples:* π and *e*.

Unit Number—the number 1.

Untouchable Numbers—numbers that are never the sum of the factors of any other number. *Examples:* 2, 5, 52, 88, 96, and 120.

Weird Numbers—a special type of abundant number that does not represent the sum of any of its factors. *Examples:* 70; 836; 4,030; 5,830; and 7,192.

LIST 11

Multiplication Table

Even in this day of high-speed computers and calculators, a basic knowledge of multiplication facts is important to understand some of the relationships between numbers. Knowing the multiplication tables is necessary for any problem that requires multiplying or dividing, especially when a calculator is not handy. Here is one example.

×	1	2	3	4	5	6	7	8	9	10	11	12
1	1	2	3	4	5	6	7	8	9	10	11	12
2	2	4	6	8	10	12	14	16	18	20	22	24
3	3	6	9	12	15	18	21	24	27	30	33	36
4	4	8	12	16	20	24	28	32	36	40	44	48
5	5	10	15	20	25	30	35	40	45	50	55	60
6	6	12	18	24	30	36	42	48	54	60	66	72
7	7	14	21	28	35	42	49	56	63	70	77	84
8	8	16	24	32	40	48	56	64	72	80	88	96
9	9	18	27	36	45	54	63	72	81	90	99	108
10	10	20	30	40	50	60	70	80	90	100	110	120
11	11	22	33	44	55	66	77	88	99	110	121	132
12	12	24	36	48	60	72	84	96	108	120	132	144

Copyright © 2005 by Judith A. Muschla and Gary Robert Muschla

LIST **12**

Rules for Finding Divisibility

Understanding divisibility is useful in many mathematical applications. Two of the most obvious are reducing fractions and finding common denominators. In advanced mathematics, divisibility tests and common factors are useful in factoring polynomials.

For a number to be divisible by

2, the number must be an even number, ending in 2, 4, 6, 8, 0.

3, the sum of the digits of the number must be divisible by 3.

4, the number must be even and the last two digits of the number are divisible by 4.

5, the number must end in 0 or 5.

6, the number must be even and the sum of its digits must be divisible by 3.

7, you must be able to drop the ones digit, and subtract two times the ones digit from the remaining number. If that answer can be divided by 7, the original number can be divided by 7.

8, the number formed by the last 3 digits of the number can be divided by 8.

9, the sum of the digits must be divisible by 9.

10, the number must end in 0.

11, you must first add the alternate digits, beginning with the first digit. Next you must add the alternate digits, beginning with the second digit. Subtract the smaller sum from the larger. If the difference is divisible by 11, the original number is divisible by 11.

12, the number must be divisible by both 3 and 4.

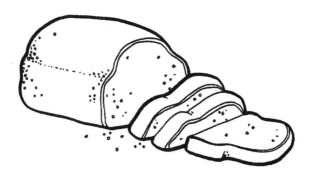

LIST 13

Rules for Finding the
Greatest Common Factor (GCF)

A *factor* is a number that divides into a larger number evenly. The *greatest common factor* (GCF) is the largest number that divides into two or more numbers evenly. Being able to find the greatest common factor is an important skill for reducing fractions. It is also helpful in algebra for factoring polynomials. The GCF of two or more numbers can be found in three ways.

Listing Factors

▶ List all the numbers that divide evenly into the first given number. These are its factors.

▶ List all the factors of the second given number.

▶ Circle the largest factor that appears in both lists. This is the greatest common factor.

Find the greatest common factor of 24 and 36.

The factors of 24 are 1, 2, 3, 4, 6, 8, ⑫, 24.

The factors of 36 are 1, 2, 3, 4, 6, 9, ⑫, 18, 36.

12 is the greatest common factor of 24 and 36.

Expressing Each Number as the Product of Primes

▶ Factor each number into its primes.

▶ Circle those factors (by pairs) common to each.

▶ The greatest common factor is the product of the numbers that are circled.

Find the greatest common factor of 24 and 36.

$$24 = 2 \times 2 \times 2 \times 3$$
$$36 = 2 \times 2 \times 3 \times 3$$
$$2 \times 2 \times 3 = 12$$

12 is the greatest common factor of 24 and 36.

Using Euclid's Method

▶ Divide the larger number by the smaller, expressing the quotient with a remainder.

▶ Divide the previous divisor by the previous remainder.

▶ Continue this process until the remainder is 0.

▶ The last divisor is the greatest common factor of the number.

Find the greatest common factor of 27 and 90.

$$\begin{array}{r} 3r9 \\ 27\overline{)90} \\ 81 \\ \hline 9 \end{array}$$

$$\begin{array}{r} 3r0 \\ 9\overline{)27} \\ 27 \\ \hline 0 \end{array}$$

9 is the greatest common factor of 27 and 90.

LIST 14

Rules for Finding the
Least Common Multiple (LCM)

A *common multiple* is a number that two other numbers will divide into evenly. The *least common multiple* (LCM) is the lowest multiple of two numbers. It is most useful for finding common denominators. The LCM of two numbers can be found in three ways.

Strategy One

▶ Start with the bigger number.

▶ List its multiples by multiplying the number by 1, 2, 3, 4, 5, etc.

▶ After each multiplication, check to see if the multiple of the larger number is also a multiple of the smaller number. If it is, you have found the least common multiple.

Find the least common multiple of 10 and 15.

15 is the larger number.
Multiples of 15: 15, 30, . . .
Is 15 a multiple of 10? No.
Is 30 a multiple of 10? Yes.
30 is the least common multiple of 10 and 15.

Strategy Two

▶ Find the product of the two numbers.

▶ Divide this product by the greatest common factor of the numbers.

Find the least common multiple of 10 and 15.

The product of 10 and 15 is 150.
The greatest common factor of 10 and 15 is 5.

$150 \div 5 = 30$. 30 is the least common multiple of 10 and 15.

Strategy Three

▶ Factor each number into its primes.

▶ Write each product using exponents.

▶ Write each base.

▶ If the base is a factor of only one number, write the base and the exponent in exponential form.

▶ If the base is a factor of more than one number, write the base in exponential form using the larger (or largest) exponents of the bases. If the exponents of a given base are the same, write the base and exponent in exponential form.

▶ The least common multiple is the product of these numbers.

Find the least common multiple of 24 and 36.

$24 = 2 \times 2 \times 2 \times 3$

$36 = 2 \times 2 \times 3 \times 3$

$24 = 2^3 \times 3$

$36 = 2^2 \times 3^2$

The bases are 2 and 3.

$2^3 \times 3^2$

$2^3 \times 3^2 = 72$
The least common multiple of 24 and 36 is 72.

LIST 15

Types of Fractions

A *fraction* is a part of a whole. There are many types of fractions.

Simple Fraction—a fraction in which the numerator and denominator are both integers. Also known as a *common fraction*.

Examples: $\frac{2}{3}$, $\frac{7}{3}$, $-\frac{6}{7}$, $\frac{5}{1}$

Proper Fraction—a fraction in which the numerator is less than the denominator.

Examples: $\frac{1}{4}$, $\frac{2}{7}$, $-\frac{1}{8}$

Improper Fraction—a fraction in which the numerator is equal to or greater than the denominator. Improper fractions are usually changed to whole or mixed numbers.

Examples: $\frac{5}{3}$, $\frac{7}{7}$, $-\frac{11}{8}$

Simplified Fraction—a fraction whose numerator and denominator are integers and their greatest common factor is 1.

Examples: $\frac{1}{2}$, $\frac{4}{7}$, $-\frac{8}{11}$

Mixed Number—a number that is a combination of an integer and a proper fraction. Thus, it is "mixed."

Examples: $2\frac{2}{3}$, $5\frac{7}{8}$, $-2\frac{1}{2}$

Unit Fraction—a fraction in which the numerator is 1.

Examples: $\frac{1}{5}$, $\frac{1}{14}$

An Integer Represented as a Fraction—a fraction in which the denominator is 1.

Examples: $\frac{2}{1}$, $-\frac{3}{1}$

Complex Fraction—a fraction in which the numerator or the denominator, or both numerator and denominator, are fractions.

Examples: $\frac{\frac{3}{5}}{\frac{7}{8}}$, $\frac{\frac{7}{9}}{4}$, $\frac{5}{\frac{1}{3}}$

Reciprocal—the fraction that results from interchanging the numerator and denominator.

Example: 4 is the reciprocal of $\frac{1}{4}$.

LIST 15

(Continued)

Similar Fractions—two or more simple fractions that have the same denominator.

Examples: $\frac{3}{7}$, $\frac{4}{7}$, $\frac{6}{7}$

Zero Fraction—a fraction in which the numerator is zero. A zero fraction equals zero.

Example: $\frac{0}{3} = 0$

Undefined Fraction—a fraction with a denominator of zero. ($\frac{7}{0}$ means 7 divided by 0, which is an impossibility because nothing can be divided by 0. Therefore, the fraction remains undefined.)

Indeterminate Form—an expression having no quantitative meaning.

Example: $\frac{0}{0}$

LIST 16

Rules for Simplifying Fractions

You can simplify fractions by dividing numerators and denominators by common factors.

▶ Find the largest number that will divide into the numerator and denominator of the fraction evenly. This number is the greatest common factor between the numerator and denominator.

Simplify $\frac{18}{24}$

6 is the greatest common factor of 18 and 24.

▶ Divide both numerator and denominator by the greatest common factor. (If you did not find the greatest common factor the first time, you can simplify further by finding another common factor.)

$\frac{18 \div 6}{24 \div 6} = \frac{3}{4}$

or

$\frac{18 \div 2}{24 \div 2} = \frac{9}{12}$ $\frac{9 \div 3}{12 \div 3} = \frac{3}{4}$

$\frac{18}{24} = \frac{3}{4}$

For Renaming Improper Fractions as Mixed Numbers

▶ Divide the numerator by the denominator.

Simplify $\frac{7}{5}$ $\frac{7}{5} = 1\frac{2}{5}$

▶ Write the mixed number. If there is no remainder, you will write a whole number.

Simplify $\frac{8}{4}$ $\frac{8}{4} = 2$

▶ Simplify the remaining fraction according to the rules above.

Simplify $\frac{18}{4}$ $\frac{18}{4} = 4\frac{2}{4} = 4\frac{1}{2}$

LIST 17

Rules for Operations with Fractions

The following rules apply to addition, subtraction, multiplication, and division of fractions.

Adding Fractions (Like Denominators)

▶ Add the numerators.

▶ Write the sum over the common denominator. (Do not add the denominators.)

▶ Simplify if possible.

$$\begin{array}{r} \frac{3}{4} \\ +\frac{2}{4} \\ \hline \frac{5}{4} = 1\frac{1}{4} \end{array}$$

Subtracting Fractions (Like Denominators)

▶ Subtract the numerators.

▶ Write the difference over the common denominator. (Do not subtract the denominators.)

▶ Simplify if possible.

$$\begin{array}{r} \frac{5}{6} \\ -\frac{1}{6} \\ \hline \frac{4}{6} = \frac{2}{3} \end{array}$$

Adding Fractions (Unlike Denominators)

▶ Find the least common denominator by finding the least common multiple of the denominators. In the example to the right, the least common multiple of the denominators is 20. Therefore, 20 is the least common denominator.

$$\begin{array}{r} \frac{4}{5} = \overline{20} \\ +\frac{3}{4} = \overline{20} \end{array}$$

▶ Write equivalent fractions with the common denominator.

▶ Add the numerators. (Do not add the denominators.)

▶ Simplify if possible.

$$\begin{array}{r} \frac{4}{5} \times \frac{4}{4} = \frac{16}{20} \\ +\frac{3}{4} \times \frac{5}{5} = \frac{15}{20} \\ \hline \frac{31}{20} = 1\frac{11}{20} \end{array}$$

LIST 17

(Continued)

Subtracting Fractions (Unlike Denominators)

▶ Find the least common denominator by finding the least common multiple of the denominators.

▶ Write equivalent fractions with the common denominator.

▶ Subtract the numerators. (Do not subtract the denominators.)

▶ Simplify if possible.

$$\frac{4}{5} = \frac{}{20}$$

$$-\frac{3}{4} = \frac{}{20}$$

$$\frac{4}{5} \times \frac{4}{4} = \frac{16}{20}$$

$$-\frac{3}{4} \times \frac{5}{5} = \frac{15}{20}$$

$$\frac{1}{20}$$

Multiplying Fractions

▶ Multiply the numerators.

▶ Multiply the denominators.

▶ Simplify if possible.

▶ For some fractions you may be able to simplify before multiplying; then follow the previous steps.

$$\frac{1}{3} \times \frac{3}{4} = \frac{3}{12} = \frac{1}{4}$$

$$\frac{\overset{1}{3}}{\underset{2}{4}} \times \frac{\overset{1}{2}}{\underset{1}{3}} = \frac{1}{2}$$

Dividing Fractions

▶ Write the first fraction.

▶ Change the division sign to a multiplication sign.

▶ Write the reciprocal of the divisor. (This is the fraction to the right of the division sign.)

▶ Multiply the numerators.

▶ Multiply the denominators.

▶ Simplify if possible.

$$\frac{3}{4} \div \frac{2}{3}$$

The reciprocal of $\frac{2}{3}$ is $\frac{3}{2}$.

$$\frac{3}{4} \div \frac{2}{3} = \frac{3}{4} \times \frac{3}{2} = \frac{9}{8} = 1\frac{1}{8}$$

LIST 18
Rules for Operations with Mixed Numbers

A *mixed number* is just what its name implies: a whole number combined with a fraction. The two are mixed together to express a value that lies somewhere between two whole numbers. Although working with mixed numbers is much like working with fractions, there are a few additional skills you will need to know.

Adding Mixed Numbers (Like Denominators)

▶ Add the numerators of the fractions. (Do not add the denominators.)

▶ Add the whole numbers.

▶ Simplify if possible.

$$3\frac{1}{5}$$
$$+2\frac{3}{5}$$
$$5\frac{4}{5}$$

Subtracting Mixed Numbers (Like Denominators) without Regrouping

▶ If the numerator of the fraction following the subtraction sign is smaller than the numerator of the first fraction, subtract the numerators. (Do not subtract the denominators.)

▶ Subtract the whole numbers.

▶ Simplify if possible.

$$5\frac{6}{7}$$
$$-3\frac{2}{7}$$
$$2\frac{4}{7}$$

Subtracting Mixed Numbers (Like Denominators) with Regrouping

▶ If the numerator following the subtraction sign is larger than the first numerator, you must regroup "1" from the whole number of the first fraction. Rewrite the "1" as a fraction with the same denominator, and add it to the first fraction.

▶ Subtract the numerators. (Do not subtract the denominators.)

▶ Subtract the whole numbers.

▶ Simplify if possible.

$$8\frac{1}{4} \qquad 8 = 7\frac{4}{4}$$
$$-3\frac{3}{4}$$

$$8\frac{1}{4} = 7\frac{4}{4} + \frac{1}{4} = 7\frac{5}{4}$$
$$-3\frac{3}{4} \qquad\qquad = 3\frac{3}{4}$$
$$4\frac{2}{4} = 4\frac{1}{2}$$

Adding Mixed Numbers (Unlike Denominators)

▶ Write equivalent fractions with the same denominators.

▶ Add the numerators. (Do not add the denominators.)

▶ Add the whole numbers.

▶ Simplify if possible.

$$3\frac{1}{3} = 3\frac{5}{15}$$
$$+2\frac{4}{5} = 2\frac{12}{15}$$
$$5\frac{17}{15} = 6\frac{2}{15}$$

LIST 18

(Continued)

Subtracting Mixed Numbers (Unlike Denominators) Without Regrouping

▶ Write equivalent fractions with the same denominators.

▶ Subtract the numerators. (Do not subtract the denominators.)

▶ Subtract the whole numbers.

▶ Simplify if possible.

$$3\frac{4}{5} = 3\frac{8}{10}$$
$$-2\frac{1}{10} = 2\frac{1}{10}$$
$$1\frac{7}{10}$$

Subtracting Mixed Numbers (Unlike Denominators) with Regrouping

▶ Write equivalent fractions with the same denominators.

▶ If necessary, rename "1" from the whole number of the first fraction and rewrite it as a fraction with the same denominator. Add it to the first fraction.

▶ Subtract the numerators. (Do not subtract the denominators.)

▶ Subtract the whole numbers.

▶ Simplify if possible.

$$4\frac{1}{8} = 4\frac{3}{24}$$
$$-2\frac{2}{3} = 2\frac{16}{24}$$

$$4 = 3\frac{24}{24}$$

$$4\frac{3}{24} = 3\frac{24}{24} + \frac{3}{24} = 3\frac{27}{24}$$
$$-2\frac{16}{24} \qquad = 2\frac{16}{24}$$
$$1\frac{11}{24}$$

Multiplying Mixed Numbers

▶ Change the mixed numbers to improper fractions. (To change a mixed number to an improper fraction, multiply the denominator by the whole number and add the numerator. Write this number over the original denominator.)

▶ If possible, simplify the fractions before multiplying.

▶ Multiply the numerators.

▶ Multiply the denominators.

▶ Simplify if possible.

$$3\frac{1}{2} \times 2\frac{4}{7} = \frac{7}{2} \times \frac{18}{7} =$$

$$\frac{\overset{1}{\cancel{7}}}{\underset{1}{\cancel{2}}} \times \frac{\overset{9}{\cancel{18}}}{\underset{1}{\cancel{7}}} = \frac{9}{1} = 9$$

LIST 18

(Continued)

Dividing Mixed Numbers

▶ Change the mixed numbers to improper fractions. (To change a mixed number to an improper fraction, multiply the denominator by the whole number and add the numerator. Write this number over the original denominator.)

$$4\frac{2}{3} \div 1\frac{1}{3} = \frac{14}{3} \div \frac{4}{3} =$$

▶ Change the divisor to its reciprocal, and rewrite the division sign as multiplication.

$$\frac{14}{3} \times \frac{3}{4} =$$

▶ If possible, simplify the fractions before multiplying.

▶ Multiply the numerators.

▶ Multiply the denominators.

$$\frac{\overset{7}{\cancel{14}}}{\underset{1}{\cancel{3}}} \times \frac{\overset{1}{\cancel{3}}}{\underset{2}{\cancel{4}}} = \frac{7}{2} = 3\frac{1}{2}$$

▶ Simplify if possible.

LIST 19

Place Value Chart

Understanding place value causes some students (and their teachers) many headaches. Since the value of any digit depends on its "place," understanding place value is an essential skill. In the example below, 5 represents 5 ten thousands and also 5 ten-millionths. The digits are the same, but the values are quite different.

trillions	hundred billions	ten billions	billions	hundred millions	ten millions	millions	hundred thousands	ten thousands	thousands	hundreds	tens	ones	.	tenths	hundredths	thousandths	ten-thousandths	hundred-thousandths	millionths	ten-millionths
3,	2	8	7,	3	8	4,	6	5	1,	2	9	6	.	3	7	8	2	6	1	5

Types of Decimals

In the broadest sense, a *decimal* is any numeral in the base ten number system. Following are several types of decimals.

Decimal Fraction—a number that has no digits other than zeros to the left of the decimal point.

   Examples: 0.349,   0.84,   0.3001

Mixed Decimal—an integer and a decimal fraction.

   Examples: 8.341,   27.1,   341.07

Similar Decimals—decimals that have the same number of places to the right of the decimal point.

   Examples: 3.87 and 0.12;   14.015 and 3.396

Decimal Equivalent of a Proper Fraction—the decimal fraction that equals the proper fraction.

   Examples: $0.25 = \frac{1}{4}$, $0.3 = \frac{3}{10}$

Finite (or Terminating) Decimal—a decimal that has a finite number of digits.

   Examples: 0.3,   0.2765,   5.38412

Infinite (or Nonterminating) Decimal—a decimal that has an unending number of digits to the right of the decimal point.

   Examples: π, $\sqrt{3}$, $0.\overline{33}$, $0.\overline{37}$,   34.12794...

Repeating (or Periodic) Decimal—a nonterminating decimal in which the same digit or group of digits repeats. A bar is used to show that a digit or group of digits repeats. The repeating set is called the period or repetend.

   Examples: $0.\overline{3}$, $0.\overline{37}$

Nonrepeating (or Nonperiodic) Decimal—a decimal that is nonterminating and nonrepeating. Such decimals are irrational numbers.

   Examples: π, $\sqrt{3}$

LIST 21

Rules for Operations with Decimals

Adding, subtracting, multiplying, and dividing decimals is not so hard as it may seem. Use the following guides.

Adding Decimals

► Line up the numbers according to the decimal points before adding. Keep columns straight and the digits in their proper places.

► Add zeros for placeholders if necessary.

► After setting up the problem, bring the decimal point straight down.

► Remember that a whole number is placed to the left of the decimal point. For example, the whole number 5 is written as 5.0.

► Add as you would with whole numbers.

► If you carry, carry to the next place.

$$2.73 + 0.145 =$$

$$\begin{array}{r} 2.73 \\ +0.145 \\ \hline \end{array}$$

$$\begin{array}{r} 2.730 \\ +0.145 \\ \hline 2.875 \end{array}$$

$$7.4 + 5 =$$

$$\begin{array}{r} 7.4 \\ +5.0 \\ \hline 12.4 \end{array}$$

Subtracting Decimals

► Line up the numbers according to decimal points before subtracting. Keep the columns straight and the digits in their proper places.

► Add zeros for placeholders if necessary.

► After setting up the problem, bring the decimal point straight down.

► Remember that a whole number is placed to the left of the decimal point. The whole number 8 is written as 8.0.

► Regroup as you would with whole numbers.

► Subtract as you would with whole numbers.

$$2.75 - 0.042 =$$

$$\begin{array}{r} 2.75 \\ -0.042 \\ \hline \end{array}$$

$$\begin{array}{r} 2.750 \\ -0.042 \\ \hline 2.708 \end{array}$$

$$9.34 - 8 =$$

$$\begin{array}{r} 9.34 \\ -8.00 \\ \hline 1.34 \end{array}$$

Multiplying Decimals

► Line up the numbers by columns, not according to decimal points.

► Multiply as you would with whole numbers.

► Count the places held by digits to the right of the decimal points in the numbers you multiplied.

► Start at the right of your answer, and count the same number of places to the left. Mark the decimal point there.

$$4.32 \times 0.7 =$$

$$\begin{array}{r} 4.32 \quad (2 \text{ places}) \\ \times \ 0.7 \ + (1 \text{ place}) \\ \hline 3.024 \quad (3 \text{ places}) \end{array}$$

LIST 21

(Continued)

Dividing a Decimal by a Whole Number

▶ Bring the decimal point straight up.
▶ Divide as you would with whole numbers.

$$
\begin{array}{r}
0.41 \\
8\,\overline{)3.28} \\
\underline{32} \\
8 \\
\underline{8} \\
\end{array}
$$

Dividing a Decimal by a Decimal

▶ Move the decimal point to the right of the divisor, making the divisor a whole number. (This is the same as multiplying a decimal by a power of 10.)

$$
\begin{array}{r}
4.1 \\
8\,\overline{)3.28} \\
\end{array}
$$

▶ Move the decimal point in the dividend to the right the same number of places.

$$
\begin{array}{r}
410. \\
8\,\overline{)328.0} \\
\underline{32} \\
8 \\
\underline{8} \\
0 \\
\underline{0} \\
\end{array}
$$
Add zero as a placeholder.

▶ Bring the decimal point straight up.
▶ Divide as you would with whole numbers.

$$
\begin{array}{r}
2.925 \\
8\,\overline{)2.3400} \\
\underline{16} \\
74 \\
\underline{72} \\
20 \\
\underline{16} \\
40 \\
\underline{40} \\
\end{array}
$$
Add zeros to finish dividing.

▶ If necessary, add a zero or zeros to the dividend to finish dividing. (The problem might work out evenly, or you might need to round off your answer.)

LIST 22

Rules for Changing Decimals to Fractions

Decimals can easily be converted to fractions. The key is understanding the place value of the decimal.

▶ Read the decimal. Here are some examples:

- One place to the right of the decimal point is tenths.
- Two places to the right of the decimal point are hundredths.
- Three places to the right of the decimal point are thousandths.
- Four places to the right of the decimal point are ten-thousandths.

▶ Write the decimal as a fraction with a denominator that is the same value of the decimal.

▶ Simplify if possible.

Change each decimal listed below to a fraction.

$$0.5 = \frac{5}{10} = \frac{1}{2}$$

$$0.23 = \frac{23}{100}$$

$$0.145 = \frac{145}{1000} = \frac{29}{200}$$

$$0.7625 = \frac{7625}{10000} = \frac{1525}{2000} = \frac{305}{400} = \frac{61}{80}$$

If the Decimal Is a Mixed Decimal

▶ Multiply by $\frac{1}{10}, \frac{1}{100}, \frac{1}{1000}$, etc. Here are some examples:

- Multiply by $\frac{1}{10}$ if the decimal has only one digit to the right of the decimal point.
- Multiply by $\frac{1}{100}$ if the decimal has two digits to the right of the decimal point.
- Multiply by $\frac{1}{1000}$ if the decimal has three digits to the right of the decimal point.

▶ Rewrite the mixed number as an improper fraction.

▶ If possible, simplify, either before or after you multiply.

Change each mixed decimal listed below to a fraction.

$$0.3\frac{1}{3} = 3\frac{1}{3} \times \frac{1}{10} = \frac{\cancel{10}^{1}}{3} \times \frac{1}{\cancel{10}_{1}} = \frac{1}{3}$$

$$0.87\frac{1}{2} = 87\frac{1}{2} \times \frac{1}{100} = \frac{\cancel{175}^{7}}{2} \times \frac{1}{\cancel{100}_{4}} = \frac{7}{8}$$

$$0.666\frac{2}{3} = \frac{\cancel{2000}^{2}}{3} \times \frac{1}{\cancel{1000}_{1}} = \frac{2}{3}$$

LIST 22

(Continued)

If the Decimal Repeats

- ▶ Multiply by 10, 100, 1,000, etc. Here are some examples:

  - Multiply by 10 if one digit repeats.

  - Multiply by 100 if two digits repeat.

  - Multiply by 1000 if three digits repeat.

- ▶ Subtract.

- ▶ Divide by the coefficient of variable.

- ▶ Simplify if possible.

Change each repeating decimal to a fraction. $0.\overline{3}$, $0.\overline{34}$, $0.\overline{371}$

$n = 0.\overline{3}$ so $10n = 3.\overline{3}$

$n = 0.\overline{34}$ so $100n = 34.\overline{34}$

$n = 0.\overline{371}$ so $1000n = 371.\overline{371}$

$$
\begin{array}{lll}
10n = 3.\overline{3} & 100n = 34.\overline{34} & 1000n = 371.\overline{371} \\
-n = -0.\overline{3} & -n = -0.\overline{34} & -n = -0.\overline{371} \\
\dfrac{9n}{9} = \dfrac{3}{9} & \dfrac{99n}{99} = \dfrac{34}{99} & \dfrac{999n}{999} = \dfrac{371}{999} \\
n = \dfrac{1}{3} & n = \dfrac{34}{99} & n = \dfrac{371}{999}
\end{array}
$$

Therefore: $0.\overline{3} = \dfrac{1}{3}$, $0.\overline{34} = \dfrac{34}{99}$, $0.\overline{371} = \dfrac{371}{999}$

LIST 23

Rules for Changing Fractions to Decimals

There are two methods for changing fractions to decimals. The first is to rewrite the fraction to make the denominator a decimal equivalent in the form of tenths, hundredths, or thousandths. The second is to divide the denominator of the fraction into its numerator. Both methods are detailed below.

Rewriting the Fraction

▶ Multiply the numerator and the denominator of the fraction so that the denominator is equal to tenths, hundredths, or thousandths.

▶ A simple way to see if this method will work is to divide the denominator into tenths, hundredths, or thousandths. If the denominator divides evenly, multiply the numerator and denominator by the same number to find the equivalent fraction.

▶ Change the fraction to an equivalent fraction.

Change $\frac{1}{2}$ to a decimal.

$$\frac{1}{2} \times \frac{5}{5} = \frac{5}{10} = 0.5$$

$$\frac{1}{2} = 0.5$$

Change $\frac{3}{4}$ to a decimal.

$$\frac{3}{4} \times \frac{25}{25} = \frac{75}{100} = 0.75$$

$$\frac{3}{4} = 0.75$$

Dividing the Numerator by the Denominator

▶ For fractions whose denominators are not equivalent to tenths, hundredths, or thousandths, divide the numerator by the denominator.

▶ Add a decimal point after the numerator, and add two zeros. (Add more zeros if you are trying to find repeating decimals. For repeating decimals, be sure to indicate the digits that repeat by putting a bar over them. In this case omit the next step.)

▶ Write the remainder as a fraction.

Change $\frac{1}{3}$ to a decimal.

$3\overline{)1}$

$$0.33\frac{1}{3} = 0.\overline{3}$$
$$3\overline{)1.00}$$
$$\underline{9}$$
$$10$$
$$\underline{9}$$
$$1$$

$$\frac{1}{3} = 0.\overline{3}$$

Rules for Changing Decimals to Percents

Percent means hundredths and is denoted by the % symbol. Changing a decimal to a percent requires that you change the decimal to hundredths first. You may do this in one of two ways: (1) change the decimal to an equivalent fraction or (2) change the decimal to a percent directly. Both methods are shown below.

Changing the Decimal to a Fraction, Then to a Percent

▶ Write the decimal as a fraction.

▶ If necessary, change the fraction to an equivalent fraction with a denominator of 100.

▶ Change the fraction to a percent.

Change 0.7 to a percent.

$$0.7 = \frac{7}{10}$$

$$\frac{7}{10} = \frac{70}{100} = 70\%$$

Changing the Decimal Directly to a Percent

▶ Move the decimal point two places to the right and include the percent symbol. (Writing the % sign indicates that the decimal was multiplied by 100.)

Change 0.7 to a percent.

$$0.70 = 70\%$$

LIST 25

Rules for Changing Percents to Decimals

Since *percent* means hundredths, percents can be converted directly to decimals. You can also convert percents to decimals by writing equivalent fractions. Both methods are shown below.

Changing the Percent Directly to a Decimal

▶ Change the percent directly to a decimal by moving the decimal point two places to the left. (This is the same as dividing by 100.)

Change each percent listed below to a decimal.

$$58\% = 0.58$$
$$125\% = 1.25$$
$$2\% = 0.02$$
$$33\tfrac{1}{3}\% = 0.33\tfrac{1}{3}$$

Changing the Percent Directly to a Decimal by Writing an Equivalent Fraction

▶ Write the percent as a fraction with a denominator of 100.

▶ Express the fraction as a decimal.

Change each percent listed below to a decimal.

$$29\% = \frac{29}{100} = 0.29$$
$$150\% = \frac{150}{100} = 1.5$$
$$33\tfrac{1}{3}\% = \frac{33\tfrac{1}{3}}{100} = 0.33\tfrac{1}{3}$$

Rules for Changing Fractions to Percents

There are two methods for changing fractions to percents. The first is to change the fraction to an equivalent fraction with a denominator of 100. The second, used for fractions that cannot be changed to equivalent fractions with denominators of 100, is to change the fraction first to a decimal, and then change the decimal to a percent.

Changing Fractions to Percents Using Equivalent Fractions

▶ If the denominator of the fraction is a factor of 100, change the fraction to an equivalent fraction with a denominator of 100. Do that by multiplying the numerator and denominator by the same number.

▶ Change the new fraction to a percent.

Change each fraction to a percent.

$$\frac{2}{5} \times \frac{20}{20} = \frac{40}{100} = 40\%$$

$$\frac{3}{4} \times \frac{25}{25} = \frac{75}{100} = 75\%$$

Changing Fractions to Percents Using the Decimal Method

▶ For fractions whose denominators are not factors of 100, divide the numerator of the fraction by its denominator.

▶ Add a decimal point and two zeros. (The two zeros are necessary to change the fraction to a percent, because percent means hundredths.)

▶ Divide. Write any remainder as a fraction.

▶ Change the decimal to a percent.

Change $\frac{4}{9}$ to a percent.

$$9\overline{)4.00}$$

$$0.44\frac{4}{9} = 44\frac{4}{9}\%$$

$$9\overline{)4.00}$$
$$\frac{36}{40}$$
$$\frac{36}{4}$$

LIST 27

Rules for Changing Percents to Fractions

The word *percent* means hundredths. $n\%$ means $n \times \frac{1}{100}$ or $n \times 0.01$. When changing percents to fractions, use the meaning of percent expressed as a fraction.

Percents can easily be changed to fractions, as the following two methods show.

Changing the Percent Directly to a Fraction

▶ Change the percent directly to a fraction with a denominator of 100. (The number of the percent becomes the numerator of the fraction.)

▶ Simplify if possible.

$$50\% = \frac{50}{100} = \frac{1}{2}$$

$$125\% = \frac{125}{100} = 1\frac{25}{100} = 1\frac{1}{4}$$

Changing the Percent When It Is a Mixed Number

▶ Multiply the percent by $\frac{1}{100}$.

▶ Simply if possible, either before you multiply or after.

Change $87\frac{1}{2}\%$ to a fraction.

$$87\frac{1}{2}\% = 87\frac{1}{2} \times \frac{1}{100} = \frac{175}{2} \times \frac{1}{100}$$

$$\frac{175^{7}}{2} \times \frac{1}{100_{4}} = \frac{7}{8}$$

$$87\frac{1}{2}\% = \frac{7}{8}$$

Percent Equivalents

The following chart shows the relationships between fractions, decimals, and percents.

Word Name	Fraction	Decimal	Percent
One-half	$\frac{1}{2}$	0.50	50%
One-fourth	$\frac{1}{4}$	0.25	25%
Three-fourths	$\frac{3}{4}$	0.75	75%
One-third	$\frac{1}{3}$	$0.33\frac{1}{3}$ or $0.\overline{3}$	$33\frac{1}{3}\%$ or $33.\overline{3}\%$
Two-thirds	$\frac{2}{3}$	$0.66\frac{2}{3}$ or $0.\overline{6}$	$66\frac{2}{3}\%$ or $66.\overline{6}\%$
One-fifth	$\frac{1}{5}$	0.20	20%
Two-fifths	$\frac{2}{5}$	0.40	40%
Three-fifths	$\frac{3}{5}$	0.60	60%
Four-fifths	$\frac{4}{5}$	0.80	80%
One-sixth	$\frac{1}{6}$	$0.16\frac{2}{3}$ or $0.1\overline{6}$	$16\frac{2}{3}\%$ or $16.\overline{6}\%$
Five-sixths	$\frac{5}{6}$	$0.83\frac{1}{3}$ or $0.8\overline{3}$	$83\frac{1}{3}\%$ or $83.\overline{3}\%$
One-eighth	$\frac{1}{8}$	$0.12\frac{1}{2}$ or 0.125	$12\frac{1}{2}\%$ or 12.5%
Three-eighths	$\frac{3}{8}$	$0.37\frac{1}{2}$ or 0.375	$37\frac{1}{2}\%$ or 37.5%
Five-eighths	$\frac{5}{8}$	$0.62\frac{1}{2}$ or 0.625	$62\frac{1}{2}\%$ or 62.5%
Seven-eighths	$\frac{7}{8}$	$0.87\frac{1}{2}$ or 0.875	$87\frac{1}{2}\%$ or 87.5%
One-ninth	$\frac{1}{9}$	$0.11\frac{1}{9}$ or $0.\overline{1}$	$11\frac{1}{9}\%$ or $11.\overline{1}\%$
Two-ninths	$\frac{2}{9}$	$0.22\frac{2}{9}$ or $0.\overline{2}$	$22\frac{2}{9}\%$ or $22.\overline{2}\%$
Four-ninths	$\frac{4}{9}$	$0.44\frac{4}{9}$ or $0.\overline{4}$	$44\frac{4}{9}\%$ or $44.\overline{4}\%$
Five-ninths	$\frac{5}{9}$	$0.55\frac{5}{9}$ or $0.\overline{5}$	$55\frac{5}{9}\%$ or $55.\overline{5}\%$

LIST 28

(Continued)

Seven-ninths	$\frac{7}{9}$	$0.77\frac{7}{9}$ or $0.\overline{7}$	$77\frac{7}{9}\%$ or $77.\overline{7}\%$
Eight-ninths	$\frac{8}{9}$	$0.88\frac{8}{9}$ or $0.\overline{8}$	$88\frac{8}{9}\%$ or $88.\overline{8}\%$
One-tenth	$\frac{1}{10}$	0.10	10%
Three-tenths	$\frac{3}{10}$	0.30	30%
Seven-tenths	$\frac{7}{10}$	0.70	70%
Nine-tenths	$\frac{9}{10}$	0.90	90%
One	1	1.00	100%

LIST 29

Rules for Solving Proportions

A *proportion* is a statement that two ratios are equal. Proportions can be helpful in solving word problems, particularly those involving percents.

▶ Write the proportion.

▶ Show the cross products of the proportion.

▶ Find the products.

▶ Divide both sides of the equation by the coefficient of *n*.

Solve for *n*.

$$\frac{n}{4} = \frac{15}{6}$$

$$4 \times 15 = 6 \times n$$

$$60 = 6n$$

$$\frac{60}{6} = \frac{6n}{6}$$

$$10 = n$$

LIST 30

Rules for Finding Percents of Numbers

The percent of a number can be found in one of three ways. The three are detailed below.

The Decimal Method

▶ Change the percent to an equivalent decimal.

▶ Multiply the decimal by the number.

Find 20% of 80.

20% = 0.2

$$\begin{array}{r} 80 \\ \times 0.2 \\ \hline 16.0 \end{array}$$

20% of 80 = 16

The Fraction Method

▶ Change the percent to an equivalent fraction.

▶ Simplify if possible.

▶ Multiply the fraction by the number.

▶ Simplify if possible.

Find 20% of 80.

$$20\% = \frac{20}{100} = \frac{1}{5}$$

$$\frac{1}{\overset{}{\cancel{5}_1}} \times \frac{\overset{16}{\cancel{80}}}{1} = \frac{16}{1} = 16 \qquad 20\% \text{ of } 80 = 16$$

The Proportion Method

▶ Write a proportion using this form:

$$\frac{Part}{Base} = \frac{Percent}{100}$$

Note that in the expression, the number after the word *of* is the base.

▶ Show the cross products of the equation.

▶ Find the products.

▶ Divide both sides of the equation by the coefficient of n.

Find 20% of 80.

$$\frac{n}{80} = \frac{20}{100}$$

$$80 \times 20 = 100 \times n$$

$$1600 = 100n$$

$$\frac{1600}{100} = \frac{100n}{100}$$

$$16 = n$$

20% of 80 = 16

LIST 31

Rules for Finding Percent and the Base

Finding what percent a number is of another number and finding a number when a certain percent of it is given are confusing operations for many students. The following lists make the steps of each process clear.

Finding the Percent

▶ Write a proportion using this form:
$$\frac{Part}{Base} = \frac{Percent}{100}$$
Note that in the expression, the number after the word *of* is the base.

▶ Show the cross products of the proportion.

▶ Find the products.

▶ Divide both sides by the coefficient of *n*.

What percent of 64 is 16?

$\frac{16}{64} = \frac{n}{100}$

$16 \times 100 = 64 \times n$

$1600 = 64n$

$\frac{1600}{64} = \frac{64n}{64}$

$25 = n$

25% of 64 = 16

Finding the Base

▶ Write a proportion using this form:
$$\frac{Part}{Base} = \frac{Percent}{100}$$
Note that the phrase "what number" after the word *of* is the base.

▶ Show the cross products of the proportion.

▶ Find the products.

▶ Divide both sides by the coefficient of *n*.

15 is 25% of what number?

$\frac{15}{n} = \frac{25}{100}$

$15 \times 100 = 25 \times n$

$1500 = 25n$

$\frac{1500}{25} = \frac{25n}{25}$

$60 = n$

15 = 25% of 60

Rules for Operations with Integers

Integers include all positive and negative whole numbers, and zero. Because negative numbers are the opposites of positives, special rules are needed for working with them.

Adding Two Integers

▶ When the integers are positive, add them, and the sign remains positive.

$$^+4 + {}^+3 = {}^+7$$

▶ When the integers are negative, add the absolute values. The sign remains negative. (The absolute value of any integer is its distance from 0 on the number line. The absolute value of both $^+4$ and $^-4$ is 4.)

$$^-4 + {}^-3 =$$
$$|{}^-4| + |{}^-3| = 4 + 3 = 7$$
$$^-4 + {}^-3 = {}^-7$$

▶ When the signs of the integers are different, subtract the absolute values (the smaller from the larger), and keep the sign of the integer with the greater absolute value.

$$^-4 + {}^+9 =$$
$$|9| - |{}^-4| = 9 - 4 = 5$$
$$^-4 + {}^+9 = 5$$

▶ When the integers are opposites, their sum is 0.

$$^-4 + {}^+4 = 0$$

Adding More than Two Integers

▶ *Method One:* Work from left to right and add integers two at a time, following the rules above.

$$^-4 + ({}^+6) + ({}^-7) + ({}^+2) =$$
$$^+2 + ({}^-7) + ({}^+2) =$$
$$^-5 + ({}^+2) = {}^-3$$

▶ *Method Two:* Add all positive integers, and then add all the negative integers, following the rules above. Find the sum of your answers.

$$^-4 + ({}^+6) + ({}^-7) + ({}^+2) =$$
$$^+6 + ({}^+2) + ({}^-4) + ({}^-7) =$$
$$^+8 + ({}^-11) = {}^-3$$
$$^-4 + ({}^+6) + ({}^-7) + ({}^+2) = {}^-3$$

LIST 32

(Continued)

Subtracting Integers

▶ Rewrite the problem by using the definition of subtraction:

$$(a - b) = a + (^-b)$$

▶ Follow the rules for adding integers.

$$^+3 - (^+6) = {}^+3 + (^-6) = {}^-3$$

$$^-2 - (^-5) = {}^-2 + (^+5) = {}^+3$$

$$^+12 - (^-4) = {}^+12 + {}^+4 = {}^+16$$

Multiplying Two Integers

▶ Find the product of the absolute values of the numbers.

$$|8| = |^-8| = 8 \qquad |7| = |^-7| = 7$$

▶ Use the correct sign:

- If both integers are positive, the answer is positive.

$$8 \times 7 = 56$$

- If both integers are negative, the answer is positive.

$$^-8 \times (^-7) = 56$$

- If one number you multiplied is positive and the other is negative, the answer is negative.

$$^-8 \times 7 = {}^-56$$

- If one of the numbers is 0, the answer is 0.

$$^-8 \times 0 = 0$$

Multiplying More Than Two Integers

▶ *Method One:* Working from left to right, multiply the integers two at a time, following the rules above.

$$^-3 \times (^-7) \times 2 =$$
$$21 \times 2 = 42$$

▶ *Method Two:* Find the product of the absolute values. If all the numbers you multiply are positive, the answer is positive. If there is an odd number of negative factors, the answer is negative. If there is an even number of negative factors, the answer is positive. If any one of the factors is 0, your answer is 0.

$$|3| = |^-3| = 3$$
$$|7| = |^-7| = 7$$
$$|2| = |^-2| = 2$$
$$^-3 \times (^-7) \times 2 = 42$$

(There are two negative factors.)

$$^-3 \times 7 \times 2 = {}^-42$$

(There is one negative factor.)

$$^-3 \times (^-7) \times (^-2) = {}^-42$$

(There are three negative factors.)

$$^-3 \times (^-7) \times 0 = 0$$

LIST 32

(Continued)

Dividing Integers

▶ Find the quotient of the absolute values. $|3| = |^-3| = 3$ $|21| = |^-21| = 21$

▶ Use the correct sign:

- If both integers are positive,
  the quotient is positive. $21 \div 3 = 7$

- If both integers are negative,
  the quotient is positive. $^-21 \div (^-3) = 7$

- If the integers have different signs,
  the quotient is negative. $^-21 \div 3 = {}^-7$

▶ If 0 is divided by an integer,
  the quotient is 0. $0 \div (^-7) = 0$

▶ If an integer is divided by 0,
  the quotient is undefined. $^-7 \div 0 = \emptyset$

LIST 33

Properties of Integers

Integers have special properties. Understanding those properties can make computation easier. The Commutative Property, for example, allows you to change the order of adding or multiplying integers. The Associative Property allows you to change grouping.

In the chart below, *a, b,* and *c* are integers.

	Addition	*Multiplication*
Closure Property	$a + b$ is an integer	ab is an integer
Commutative Property	$a + b = b + a$	$ab = ba$
Associative Property	$(a + b) + c = a + (b + c)$	$(ab)c = a(bc)$
Identity Property	$a + 0 = a$	$1(a) = a$
Inverse Property	$a + {}^-a = 0$	
Multiplication Property of Zero		$a(0) = 0$
Distributive Property	$a(b + c) = ab + bc$	

LIST 34

Rules for Finding the Average (Mean)

An *average,* also referred to as the *mean,* is a number that represents a set of numbers. Averages are useful for comparing data. There are many kinds of averages, including batting averages in baseball, average incomes, and the average grade you maintain in your math class.

A *weighted average* is an average in which one item (or items) is given more importance, or "weight," than the others. For example, a unit test might count for 50% of a student's test grade, while two chapter tests count for 25% each. In this case, the unit test would be counted twice in finding the student's test average.

Steps to finding both an average and weighted average follow.

Finding an Average

▶ Add all the items you need to average.

▶ Divide the sum by the total number of items you added.

▶ If necessary, add a decimal point and zeros. (It is usually not necessary to work out problems past the hundredths place. Round off to the nearest tenth.)

Find the average of 92, 84, and 87.

$$92 + 84 + 87 = 263$$

$$3\overline{)263.00} \quad 87.66 \approx 87.7$$

Finding a Weighted Average

▶ Multiply each item by its weight. For example, if an item is to be counted twice, it will be multiplied by 2. If an item is to be counted three times, it is multiplied by 3. (An item that is not weighted is multiplied by one.)

▶ Find the sum of the items and weighted items.

  ▪ Find the sum of the weights.

  ▪ Divide the sum of the items by the sum of the weights.

  ▪ If necessary, add a decimal point and zeros, and round your answer to the desired place.

Find the average of 92, 84, and 87, if 87 is weighted twice.

$$92 \times 1 = 92$$
$$84 \times 1 = 84$$
$$87 \times 2 = 174$$

Sum of the numbers = 350

Sum of the weights = 4

$$4\overline{)350.0} \quad 87.5$$

Rules for Rounding Numbers

Rounding is an important estimation skill. When you go to the grocery store, for example, it can be helpful to round off and estimate the cost of the items you are buying before you actually buy them.

Steps for Rounding Up

▶ Circle the digit that is to be rounded.

▶ If the digit to the right is 5 or greater, round the "circled" digit up by adding 1 to it.

▶ Change all digits to the right of the rounded digit to zeros.

▶ Delete any zeros that are not placeholders.

Round 3854 to the nearest thousand.

③854 ≈ 4000

Round 2.874 to the nearest tenth.

2.⑧74 ≈ 2.9̶0̶0̶

≈ 2.9

Steps for Rounding Down

▶ Circle the digit that is to be rounded.

▶ If the digit to the right is less than 5, the "circled" digit stays the same.

▶ Change all digits to the right of the rounded digit to zeros.

▶ Delete any zeros that are not placeholders.

Round 3512 to the nearest hundred.

3⑤12 ≈ 3500

Round 2.874 to the nearest hundredth.

2.8⑦4 ≈ 2.87̶0̶

≈ 2.87

When 9 Is in the Place You Are Rounding

▶ Circle the digit that is to be rounded, which in this case is 9.

▶ If the digit to the right is 5 or greater, round the "circled" 9 up by adding 1 to it. If the digit to the right is less than 5, follow the steps for rounding down.

▶ Since 9 + 1 = 10, write 0 in place of the 9 and add 1 to the digit to the left.

▶ Change all numbers to the right of the rounded number to zeros.

▶ Delete any zeros that are not placeholders.

Round 3985 to the nearest hundred.

3⑨85 ≈ 4000

Round 2.897 to the nearest hundredth.

2.8⑨7 ≈ 2.90̶0̶

≈ 2.90

LIST 36

Rules for Finding Prime Factorizations

The *prime factorization* of a number means expressing the number as a product of prime numbers. The following lists are helpful for finding prime factorizations.

Finding the Prime Factorization through a Factor Tree

▶ Find any pair of factors of the number.

▶ Circle the prime factor(s).

▶ Find any other factors.

▶ Circle the prime factor(s).

▶ When you have found all the prime factors, write them out as a product.

▶ Write the product using exponents.

Find the prime factorization of 28.

$$\begin{array}{ccc} 28 & & 28 \\ \wedge & or & \wedge \\ ②\times 14 & & 4\times ⑦ \\ \wedge & & \wedge \\ ②\times ⑦ & & ②\times ② \end{array}$$

$$2\times 2\times 7 \qquad\qquad 2\times 2\times 7$$

$$2^2\times 7 \qquad\qquad 2^2\times 7$$

$2^2 \times 7$ is the prime factorization of 28.

Finding the Prime Factorization by Dividing by Primes

▶ Divide by 2 if possible, until the quotient is no longer divisible by 2.

▶ Divide by 3 if possible, until the quotient is no longer divisible by 3.

▶ Divide by 5 if possible, until the quotient is no longer divisible by 5.

▶ Continue this pattern, dividing by prime numbers only, until the quotient is prime.

▶ Write the product of the divisors and quotient as a product.

▶ Write the product of the divisors and quotient using exponents.

Find the prime factorization of 140.

$$2\overline{)140}^{\,70} \quad \text{Divide by 2.}$$

$$2\overline{)70}^{\,35} \quad \text{Divide by 2.}$$

$$5\overline{)35}^{\,7} \quad \begin{array}{l}\text{Cannot divide by 3; so} \\ \text{divide by 5. 7 is prime.}\end{array}$$

$$2\times 2\times 5\times 7$$

$2^2\times 5\times 7$ is the prime factorization of 140.

Scientific Notation

Scientific notation is used to express very large or very small numbers. For example, the mean distance of Pluto from the sun is about 3,670,000,000 miles. That is a jawbreaker to say and even worse to write. Using scientific notation, the number can be expressed as 3.67×10^{9}.

In writing scientific notation for large numbers, follow these rules:

▶ The first factor must be greater than or equal to 1 and less than 10.

▶ The second factor must be a power of 10 expressed in exponential form.

▶ To write numbers with exponents, count the number of places to the right of the first nonzero number in standard form. Use the number of places as the exponent. In the case of 3,670,000,000, the first nonzero number is 3. Since 9 digits are to the right of the 3, 9 is the exponent. Therefore, $3,670,000,000 = 3.67 \times 10^{9}$.

In writing very small numbers, follow these rules:

▶ The first factor must be greater than or equal to 1 and less than 10.

▶ The second factor must be a negative power of 10 expressed in exponential form.

▶ To write numbers with a negative exponent, count the number of places to the right of the decimal point, up to and *including* the first nonzero number. Use the number of places as the negative exponent. For example, $0.00079 = 7.9 \times 10^{-4}$.

LIST **38**

Bases

The *base* of any number system is the number of different symbols used. The Arabic system, which we use, is a base ten system because ten symbols are used in writing numerals: 0, 1, 2, 3, 4, 5, 6, 7, 8, 9. It is thought that the base ten system reflects our eight fingers and two thumbs. Primitive people found it easier to count that way.

Number systems can be based on any amount of symbols, however. The binary system (base two), for example, has only two digits, 0 and 1. Computers perform their calculations in binary codes. Pulses of electrical energy representing 0 and 1 turn tiny switches on and off in microcircuits.

Following is a comparison of the numerals of the base ten system with those of base two, base five, and base eight.

Base Ten	Base Two	Base Five	Base Eight
1	1	1	1
2	10	2	2
3	11	3	3
4	100	4	4
5	101	10	5
6	110	11	6
7	111	12	7
8	1,000	13	10
9	1,001	14	11
10	1,010	20	12
11	1,011	21	13
12	1,100	22	14
13	1,101	23	15
14	1,110	24	16
15	1,111	30	17
16	10,000	31	20
17	10,001	32	21
18	10,010	33	22
19	10,011	34	23
20	10,100	40	24

LIST **39**

Big Numbers

Most of us can comprehend numbers up to around a hundred thousand. That is about how many people a big football stadium can seat. If we use our imaginations and envision five stadiums that large side by side, all filled, we can grasp a half-million. Ten such filled stadiums add up to 1 million, but after that, the numbers soon become too large to imagine. The following list shows numbers up to a googolplex, which is big even for big numbers. To eliminate the appearance of a page covered with zeros, we use exponents to express numbers after 1 decillion.

Word Name	Digits
One million	1,000,000
One billion	1,000,000,000
One trillion	1,000,000,000,000
One quadrillion	1,000,000,000,000,000
One quintillion	1,000,000,000,000,000,000
One sextillion	1,000,000,000,000,000,000,000
One septillion	1,000,000,000,000,000,000,000,000
One octillion	1,000,000,000,000,000,000,000,000,000
One nonillion	1,000,000,000,000,000,000,000,000,000,000
One decillion	1,000,000,000,000,000,000,000,000,000,000,000
One undecillion	10^{36}
One duodecillion	10^{39}
One tredecillion	10^{42}
One quattuordillion	10^{45}
One quindecillion	10^{48}
One sexdecillion	10^{51}
One septendecillion	10^{54}
One octodecillion	10^{57}
One novemdecillion	10^{60}
One vigintillion	10^{63}
One googol	10^{100}
One googolplex	10^{googol}

Are numbers infinite? A simple test suggests that they are. Try imagining the biggest number you can. No matter how big it is, it is not the biggest. You can always add at least 1 to it.

LIST 40

Mathematical Signs and Symbols

The following list provides the signs and symbols used most often in mathematics.

$+$	addition, plus, positive (if the sign is on the upper left-hand side of a number)
$-$	subtraction, minus, opposite of, negative (if the sign is on the upper left-hand side of a number)
$\times$	multiplication, multiply by, times
$\bullet$	multiplication, multiply by, times
$*$	multiplication, multiply by, times
$\div$	division, divide by
$\dfrac{x}{y}$	division
$/$	division
$=$	equals, is equal to
$\approx$	is approximately equal to
$\equiv$	equivalence
$\neq$	is not equal to
$>$	is greater than
$\geq$	is greater than or equal to
$<$	is less than
$\leq$	is less than or equal to
$\therefore$	therefore
∞	infinity
$\$$	dollar sign
$¢$	cents sign
$\#$	number or pounds
$\%$	percent
$:$	is to
$\triangle$	triangle
$\square$	square
$\angle ABC$	angle ABC
$m\angle ABC$	the measure of angle ABC
$\ulcorner$	right angle
$\overset{\frown}{AB}$	arc AB
$\parallel$	is parallel to
$\perp$	is perpendicular to

LIST 40

(Continued)

$\overrightarrow{AB}$	ray AB
$\overline{AB}$	segment AB
$m\angle AB$	measure of line segment AB
$\overleftrightarrow{AB}$	line AB
π	pi, which is about 3.14
$\cong$	is congruent to
$\sim$	is similar to
$\propto$	is proportional to
( )	parentheses, grouping symbol
[ ]	braces, grouping symbol
$\pm$	plus or minus
$|n|$	absolute value of n
(x,y)	ordered pair
x^a	x to the a power
$\sqrt{}$	positive square root
$-\sqrt{}$	negative square root
$f(x)$	f of x, the value of f at x
{ }	indicates a set, or is used as a grouping symbol
ϕ	empty set
$\in$	is an element of
$\cap$	intersection
$\cup$	union
$P(E)$	probability of event E
$n!$	n factorial
nPr	number of permutations of n items, taken r at a time
nCr	number of combinations of n items, taken r at a time
θ	theta, an angle in standard position
$\mathrm{Sin}\,\theta$	sine of θ
$\mathrm{Cos}\,\theta$	cosine of θ
$\mathrm{Tan}\,\theta$	tangent of θ
$\mathrm{Cot}\,\theta$	cotangent of θ
$\mathrm{Sec}\,\theta$	secant of θ
$\mathrm{Csc}\,\theta$	cosecant of θ

Section Two

Measurement

Things That Measure

Unless they really think about it, most people do not realize how many things we measure. The following list gives some examples.

Measuring Device	*What It Measures*
Accelerometer	increase of speed
Altimeter	altitude
Ammeter	amperage
Anemometer	velocity of wind
Atomic clock	time
Audiometer	hearing
Balance	weight
Barometer	atmospheric pressure
Beaker	capacity
Calendar	days, weeks, months
Caliper	dimensions of a place
Chronometer	time (used on ships)
Clock	time
Egg timer	timer for cooking eggs
Electric meter	kilowatts used
Eyedropper	small liquid quantities
Fathometer	depth of water
Flask	capacity
Galvanometer	electric current
Gas gauge	amount of gasoline in a tank
Gas meter	quantity of gas
Gasometer	gases
Geiger counter	radiation
Hourglass	time (by grains of sand)
Hydrometer	density of a liquid
Hygrometer	relative humidity
IQ tests	intelligence
Jeweler's stick	ring size
Light meter	light (in photography)
Manometer	pressure (gases, liquids)
Measuring cup	capacity
Measuring spoon	capacity
Meter stick	length (in the metric system)

LIST 41

(Continued)

Measuring Device	What It Measures
Metronome	tempo (in music)
Micrometer	very small distances
Odometer	distance traveled
Oil gauge	oil pressure
Pedometer	number of steps taken
Platform scale	weight (heavy objects)
Potentiometer	voltage
Protractor	angles
Radiometer	radiation
Rain gauge	amount of rainfall
Ruler	length
Scale	weight
Seismograph	intensity of earthquakes
Sextant	angular distance
Snow gauge	amount of snowfall
Speedometer	speed of a vehicle
Sphygmomanometer	blood pressure
Spirometer	volume of air entering and leaving lungs
Stopwatch	short periods of time
Sundial	time of day
Tachometer	speed of rotation
Tape measure	length
Temperature gauge	temperature
Theodolite	angles (in surveying)
Thermometer	temperature
Timer	time
Tire gauge	air pressure in tires
Voltmeter	electrical force
Watch	time
Water meter	amount of water
Yardstick	length

The Origin of Measurements

When our ancestors first found the need to measure things, it was natural that they would use objects with which they were most familiar. Fingers, hands, arms, and feet were some of the earliest units for the measurement of length. In some cases, metal bars of a precise length were also used. Stones were often designated as units of weight, and baskets could easily be used for capacity. Of course, problems arose when someone's foot was bigger than another's or villages relied on different sizes of baskets to measure corn. Clearly, our units of measurement today are improvements over what our ancestors used. Following is a sampling of early units of measure.

Length

inch—the length of three barley grains placed end to end. (Sometimes corn kernels or other grains were used.)

digit—the breadth of a finger, about 0.75 inch.

palm—the breadth of a hand, about 4 inches.

hand—the length from the wrist to the end of the middle finger, about 8 inches.

cubit—the length of the forearm from the point of the elbow to the end of the middle finger, about 18 inches.

foot—in many ancient cultures the length of a man's foot, about 12 inches. In ancient Rome a foot equaled 4 palms.

fathom—the length of rope when pulled between a man's two outstretched arms. Sailors used fathoms to measure the depth of the water on which their ship sailed. The fathom, still used today, is 6 feet.

furlong—the length of a short race. Today the furlong is equal to one-eighth of a mile.

mile—the distance of a thousand paces, as marked off by a length of 5 feet between lifts of the same foot. The modern mile is 5,280 feet.

league—the distance a person can see across a flat field, about 3 miles.

Capacity and Weight

Depending on the time and place, jars, bowls, and baskets were all used to measure capacity. The measuring of weight also varied. Simple balances compared the weight of one object with that of another. Sometimes stones were designated as standards for comparing weight.

LIST 43

Obsolete Units of Measure

Just as language changes over time, so do the types of units used for measurement. It is likely that you never heard of some of the following terms.

Length	Dry Measure	Liquid Measure
3 barleycorns = 1 inch	2 quarts = 1 bottle	4 gills = 1 pint
$2\frac{1}{2}$ inches = 1 nail	2 bottles = 1 gallon	1 hogshead = 63 gallons
4 nails = 1 quarter	2 gallons = 1 peck	
4 quarters = 1 yard	3 bushels = 1 sack	
4 inches = 1 hand	4 bushels = 1 coomb	
3 inches = 1 palm	9 bushels = 1 vat	
4 digits = 1 palm	2 coombs = 1 quarter	
3 palms = 1 span	5 quarters = 1 wey	
7 palms = 1 cubit	2 weys = 1 last	
18 inches = 1 cubit		
5 feet = 1 pace		
$5\frac{1}{2}$ yards = 1 pole		
40 poles = 1 furlong		
8 furlongs = 1 mile		
3 miles = 1 league		

In time, assuming that the United States makes a total conversion to the metric system, it is quite possible that most of the English units of measurement will join these obsolete measuring units.

LIST 44

Measurement Abbreviations

The following list includes abbreviations for units in both the English system and metric system.

English System	*Metric System*
inch—in *or* "	nanometer—nm
foot—ft *or* '	millimeter—mm
yard—yd	centimeter—cm
rod—rd	decimeter—dm
furlong—fur	meter—m
mile—mi	dekameter *or* decameter—dkm *or* dam
fathom—fm	hectometer—hm
	kilometer—km
grain—gr	
pennyweight—dwt *or* pwt	milliliter—mL
ounce (troy)—oz t	centiliter—cL
pound (troy)—lb t	deciliter—dL
	liter—L
dram—dr	dekaliter *or* decaliter—dkL *or* daL
ounce—oz	hectoliter—hL
pound—lb *or* #	kiloliter—kL
hundredweight—cwt	
ton—T	milligram—mg
short ton—sh t	centigram—cg
long ton—l t	decigram—dg
	gram—g
gill—gi	dekagram *or* decagram—dkg *or* dag
pint—pt	hectogram—hg
quart—qt	kilogram—kg
gallon—gal	metric ton—t
barrel—bbl	
	square millimeter—sq mm *or* mm^2
peck—pk	square centimeter—sq cm *or* cm^2
bushel—bu	square decimeter—sq dm *or* dm^2
	square meter—sq m *or* m^2
chain—ch	square dekameter *or* square decameter—
	sq dkm *or* sq dam *or* dkm^2 *or* dam^2
cup—c	hectare *or* square hectometer—ha *or* hm^2
teaspoon—t *or* tsp	square kilometer—sq km *or* km^2
tablespoon—T *or* tbsp	
fluid ounce—fl oz	

LIST 44

(Continued)

English System	Metric System
square inch—sq in *or* in^2	cubic millimeter—cu mm *or* mm^3
square foot—sq ft *or* ft^2	cubic centimeter—cc *or* cm^3
square yard—sq yd *or* yd^2	cubic decimeter—cu dm *or* dm^3
square rod—sq rd *or* rd^2	cubic meter—cu m *or* m^3
acre—A	cubic dekameter *or* cubic decameter—
square mile—sq mi *or* mi^2	cu dkm *or* cu dam *or* dkm^3 *or* dam^3
	cubic hectometer—cu hm *or* hm^3
cubic inch—cu in *or* in^3	cubic kilometer—cu km *or* km^3
cubic foot—cu ft *or* ft^3	
cubic yard—cu yd *or* yd^3	
cord—cd	

LIST **45**

The English System of Weights and Measures

The United States is one of the few countries that still uses the English system of measurement. Much of the rest of the world relies on the metric system. Although the United States is slowly changing over to the metric system, the English system is far from being a relic yet, and many Americans would have trouble measuring anything without it.

Linear Measure (Length)

1,000 mils = 1 inch

12 inches = 1 foot

3 feet = 1 yard

5.5 yards = 1 rod

4 rods = 1 chain

10 chains = 1 furlong

40 rods = 1 furlong

8 furlongs = 1 mile

5,280 feet = 1 mile

1,760 yards = 1 mile

3 miles = 1 league

Nautical Linear Measure

6 feet = 1 fathom

120 fathoms = 1 cable's length (U.S. Navy)

1 nautical mile = 6,076.12 feet

1 nautical mile per hour = 1 knot

60 nautical miles = 1 degree of a great circle of the Earth

Surveyor's Measure

7.92 inches = 1 link

100 links = 1 chain

66 feet = 1 chain

10 chains = 1 furlong

80 chains = 1 mile

LIST 45

(Continued)

Square Measure (Area)

144 square inches = 1 square foot

9 square feet = 1 square yard

30.25 square yards = 1 square rod

160 square rods = 1 acre

4,840 square yards = 1 acre

640 acres = 1 square mile

Cubic Measure (Volume)

1,728 cubic inches = 1 cubic foot

27 cubic feet = 1 cubic yard

231 cubic inches = 1 gallon (U.S.)

277.27 cubic inches = 1 gallon (U.K.)

2,150.42 cubic inches = 1 bushel (U.S.)

2,219.36 cubic inches = 1 bushel (U.K.)

Liquid Measure (Capacity)

3 teaspoons = 1 tablespoon

2 tablespoons = 1 fluid ounce

4 fluid ounces = 1 gill

8 fluid ounces = 1 cup

2 cups = 1 pint

4 gills = 1 pint

2 pints = 1 quart

4 quarts = 1 gallon

Dry Measure (Capacity)

2 pints = 1 quart

8 quarts = 1 peck

4 pecks = 1 bushel

1 British dry quart = 1.032 U.S. dry quarts

LIST 45

(Continued)

Dry Measure (Cooking)

1 pinch = $\frac{1}{8}$ teaspoon

1 dash = $\frac{1}{16}$ teaspoon

1 sprinkle = $\frac{1}{32}$ teaspoon

Weight (Avoirdupois)

The avoirdupois system is used for general weighing.

27.3438 grains = 1 dram

16 drams = 1 ounce

16 ounces = 1 pound

14 pounds = 1 stone

4 quarters = 1 long hundredweight (U.K.)

100 pounds = 1 short hundredweight (U.S.)

2,000 pounds = 1 short ton (U.S.)

2,240 pounds = 1 long ton (U.K.)

Weight (Troy)

The troy system is used for weighing precious metals or gems.

1 carat = 3.086 grains

24 grains = 1 pennyweight

20 pennyweights = 1 ounce

12 ounces = 1 pound

Weight (Apothecaries)

The apothecaries system formerly was used by pharmacists. Today, most pharmacists rely on metric units.

20 grains = 1 scruple

3 scruples = 1 dram

8 drams = 1 ounce

12 ounces = 1 pound

LIST 45

(Continued)

Wood Measure

144 cubic inches = 1 board foot = $1' \times 1' \times 1'$

16 cubic feet = 1 cord foot = $4' \times 4' \times 1'$

8 cord feet = 1 cord

Angular or Circular Measure

60 seconds = 1 minute

60 minutes = 1 degree

30 degrees = 1 zodiac sign

57.2958 degrees = 1 radian

90 degrees = 1 quadrant or right angle

360 degrees = 1 circle

Hardness of Some Popular Gems

The scale for hardness runs from 10 to 1, with 10 being the hardest.

Diamond—10

Corundum—9

Topaz—8

Quartz—7

Labradorite—6

Smithsonite—5

Fluorite—4

Calcite—3

Alabaster—2

LIST 46
Rules for Converting Units in the English System

Because the values of units in the English system of measurement vary, unlike the metric system in which units are multiples of 10, converting one unit to another often requires using equivalencies. While this list offers guidelines for conversions, List 45, "The English System of Weights and Measures," is a good source for the units of the English system.

To Convert from a Larger Unit to a Smaller Unit

▶ Refer to an equivalency table to find a relationship between both quantities.

▶ Multiply.

▶ Add if necessary.

Convert 4 ft 8 in to inches.
1 ft = 12 in
4 ft = 4 × 12 = 48 in

4 ft 8 in = 48 in + 8 in = 56 in

To Convert from a Smaller Unit to a Larger Unit

▶ Refer to an equivalency table to find a relationship between both quantities.

▶ Divide.

▶ Express the remainder as the smaller unit of equivalency.

Convert 20 fl oz to cups and fluid ounces.
8 fl oz = 1 cup

$$
\begin{array}{r}
2r4 \\
8\overline{)20} \\
\underline{16} \\
4
\end{array}
$$

20 fl oz = 2 c 4 fl oz

LIST 47

U.S. and British Units of Measurement

Although the U.S. system of measurement is based largely on traditional British units, there are some differences. Both the U.S. and traditional British system are commonly referred to as the English system. Sometimes this can cause confusion.

Weight (Avoirdupois)

Unit	U.S. Equivalent	British Equivalent
1 ounce	437.5 grains	437.5 grains
1 pound	16 ounces	16 ounces
1 stone	none	14 pounds
1 hundredweight*	100 pounds	8 stones or 112 pounds
1 ton**	2,000 pounds	20 hundredweight or 2,240 pounds

*Hundredweight in the United States is known as a *short* hundredweight. In the traditional British system, it is a *long* hundredweight.

**The ton in the United States is the *short* ton, while in the traditional British system, it is a *long* ton.

Liquid Measure

Unit	U.S. Equivalent	British Equivalent
1 fluid ounce	1.8047 cubic inches	1.734 cubic inches
1 pint	16 fluid ounces or 28.88 cubic inches	20 (British) fluid ounces or 34.68 cubic inches
1 quart	2 pints or 57.75 cubic inches	2 (British) pints or 69.36 cubic inches
1 gallon*	4 quarts or 231 cubic inches	1 Imperial gallon** or 4 (British) quarts or 277.42 cubic inches

*1 U.S. gallon = 0.833 British Imperial gallon.

**1 British Imperial gallon = 1.201 U.S. gallons.

LIST 48

Units of the Metric System

The official name of the metric system is the International System of Units (commonly referred to throughout the world as Système International, or SI). While most people are familiar with the meter, gram, and liter, the metric system contains several more units of measurement. They are divided into three categories: the basic units, the supplementary units, and the derived units. Of the derived units, only the most common are included in this list.

Basic Units

Unit	Symbol	Quantity
meter	m	length
kilogram	kg	mass
second	s	time
ampere	A	electric current
Kelvin	K	temperature
candela	cd	luminous intensity
mole	mol	amount of substance

Supplementary Units

Unit	Symbol	Quantity
radian	rad	plane angle
steradian	sr	solid angle

Derived Units

Unit	Symbol	Quantity
square meter	m^2	area
cubic meter	m^3	volume
kilogram per cubic meter	kg/m^3	density
meter per second	m/s	velocity
meter per second squared	m/s^2	acceleration
newton	N	force
joule	J	energy
hertz	Hz	frequency (electromagnetism)
watt	W	power
volt	V	voltage
ohm	Ω	electrical resistance

LIST **49**

Metric Standards

Standards in measurement refer to the physical representations of the value of a unit of measure. The standards themselves are not used for direct measurement; they are used only for reference and ensure that units of measure remain accurate. They are so important that they are kept in vaults where temperature, humidity, and security can be controlled.

Basic Units

Meter (length)—equal to the distance traveled by light in a vacuum in $\frac{1}{299,792,458}$ second.

kilogram (mass)—the unit of mass equal to the mass of the platinum-iridium cylinder kept by the International Bureau of Weights and Measures in France. (The kilogram is 1,000 grams. The weight of a gram is equal to 1 cubic centimeter of water at its maximum density.)

Second (time)—the duration of 9,192,631,770 periods of radiation corresponding to a specific transition of the cesium-133 atom.

Ampere (electric current)—the constant current that, flowing in two parallel conductors 1 meter apart in a vacuum, will produce a force of 2×10^{-7} newtons per meter of length.

Kelvin (temperature)—based on the triple point of water, which is the point at which water can exist in three states: liquid, vapor, and ice. It is defined as $\frac{1}{273.16}$ of the temperature of the triple point of water.

Candela (luminous intensity)—intensity of the black-body radiation from a surface of $\frac{1}{600,000}$ square meter at the temperature of freezing platinum and at a pressure of 101,325 pascals.

Mole (amount of substance)—an amount of a substance in a system that contains as many elementary entities as there are atoms in 0.012 kilogram of carbon 12.

(Continued)

Supplementary Units

Following are the standards for the two supplementary units in the metric system.

Radian (plane angle)—a unit of angular measure equal to the central angle whose sides are two radii of a circle that cut off an arc whose length is equal to the radius of the circle.

Steradian (solid angle)—a unit of measure equal to the solid angle whose vertex is in the center of a sphere, which cuts off an area equal to the radius squared on the surface of the sphere.

Derived Units

Derived units are defined in terms of the basic units. In some cases, they have special names and symbols. The major derived units and their standards are listed below.

Newton—the unit of force equal to the force needed to accelerate 1 kilogram by 1 meter per second squared.

Joule—the unit of energy and work equal to the work done when the point of application of a force on 1 newton moves 1 meter in the direction of the force.

Hertz—the unit of frequency in the field of electromagnetism defined as 1 cycle per second.

Watt—the unit of power defined as the power of 1 joule per second.

Volt—the unit of voltage defined as the difference of electrical potential between two points of a conductor carrying a constant current of 1 ampere when the power used between them equals 1 watt.

Ohm—the unit of electrical resistance equal to a resistance that passes a current of 1 ampere when there is an electrical potential difference of 1 volt across it.

LIST **50**

Metric Prefixes

Most people are familiar with some of the prefixes of the metric system, most notably *kilo* (meaning one thousand), *centi* (one-hundredth), and *milli* (one-thousandth). There are other metric prefixes that describe numbers vastly bigger and incredibly smaller. For example, a megameter equals 1 million meters; a nanometer equals 1 billionth of a meter. In the following list, the prefixes range from biggest to smallest.

Metric Prefix	Symbol	Value
yotta	Y	one septillion *or* 10^{24}
zetta	Z	one sextillion *or* 10^{21}
exa	E	one quintillion *or* 10^{18}
peta	P	one quadrillion *or* 10^{15}
tera	T	one trillion *or* 10^{12}
giga	G	one billion *or* 10^{9}
mega	M	one million *or* 10^{6}
kilo	k	one thousand *or* 10^{3}
hecto	h	one hundred *or* 10^{2}
deka *or* deca	dk *or* da	ten *or* 10^{1}

The basic unit has no prefix.

deci	d	one-tenth *or* 10^{-1}
centi	c	one-hundredth *or* 10^{-2}
milli	m	one-thousandth *or* 10^{-3}
micro	μ	one-millionth *or* 10^{-6}
nano	n	one-billionth *or* 10^{-9}
pico	p	one-trillionth *or* 10^{-12}
femto	f	one-quadrillionth *or* 10^{-15}
atto	a	one-quintillionth *or* 10^{-18}
zepto	z	one-sextillionth *or* 10^{-21}
yocto	y	one-septillionth *or* 10^{-24}

LIST 51

Weights and Measures in the Metric System

In the metric system, the basic unit of length is the *meter*, the basic unit of weight is the *gram*, and the basic unit of capacity is the *liter*. To change a unit to a larger one, simply multiply by a power of 10. To change to a smaller unit, divide by a power of 10. There are few equivalences to remember; the most important items are the prefixes. The metric system is now the standard for measurement in most countries.

Linear Measure

10 millimeters = 1 centimeter

10 centimeters = 1 decimeter

10 decimeters = 1 meter

10 meters = 1 dekameter
or decameter

10 dekameters = 1 hectometer

10 hectometers = 1 kilometer

Square Measure

100 square millimeters =
1 square centimeter

100 square centimeters =
1 square decimeter

100 square decimeters =
1 square meter

100 square meters =
1 square dekameter *or*
square decameter

100 square dekameters =
1 square hectometer *or* hectare

100 square hectometers =
1 square kilometer

Liquid Measure

10 milliliters = 1 centiliter

10 centiliters = 1 deciliter

10 deciliters = 1 liter

10 liters = 1 dekaliter *or* decaliter

10 dekaliters = 1 hectoliter

10 hectoliters = 1 kiloliter

Cubic Measure

1000 cubic millimeters =
1 cubic centimeter = 1 milliliter

1000 cubic centimeters =
1 cubic decimeter = 1 liter

1000 cubic decimeters = 1 cubic meter

1000 cubic meters = 1 cubic dekameter
or cubic decameter

1000 cubic dekameters =
1 cubic hectometer

1000 cubic hectometers =
1 cubic kilometer

Mass Measure (Weight)

10 milligrams = 1 centigram

10 centigrams = 1 decigram

10 decigrams = 1 gram

10 grams = 1 dekagram *or* decagram

10 dekagrams = 1 hectogram

10 hectograms = 1 kilogram

100 kilograms = 1 quintal

10 quintals = 1 metric ton

LIST 52

Converting One Unit to Another in the Metric System

The values of metric units are based on tens. This is why the metric system is much easier to use than the English system. While List 51, "Weights and Measures in the Metric System," shows the relationships between metric units, the following steps can help you convert metric units of length, liquid capacity, and mass.

To Convert from a Larger Unit to a Smaller Unit

▶ List the units of the equivalency from greatest to least.

▶ Move the decimal point to the right every time you move from one unit to another. (Moving the decimal point to the right is the same as multiplying by 10.)

▶ Insert placeholders if necessary.

Convert 2.8 kg to dg.

kg hg dkg g dg
  1 2 3 4

2.8000∧

2.8 kg = 28,000 dg

To Convert from a Smaller Unit to a Larger Unit

▶ List the units of the equivalency from greatest to least.

▶ Move the decimal point to the left every time you move from one unit to another. (Moving the decimal point to the left is the same as dividing by 10.)

▶ Insert placeholders if necessary.

Convert 35mm to dm.

dm cm mm
  2 1

∧35

35 mm = .35 dm

LIST 53

Rules for Adding Units of Measurement

When you must add units in measurement, follow these steps.

▶ Line up the units in columns.

▶ Add each column (unit) separately.

▶ Simplify your answer if possible.

Find the sum of 3 ft 7 in and 4 ft 8 in.

$$
\begin{array}{r}
3 \text{ ft } 7 \text{ in} \\
+ \, 4 \text{ ft } 8 \text{ in} \\
\hline
7 \text{ ft } 15 \text{ in}
\end{array}
$$

1 ft = 12 in

$$
\begin{array}{r}
1r3 \\
12\overline{)15} \\
\underline{12} \\
3
\end{array}
$$

15 in = 1 ft 3 in

7 ft 15 in = 7 ft + 1 ft 3 in = 8 ft 3 in

LIST 54

Rules for Subtracting Units of Measurement

Use the following steps for subtracting units of measurement.

▶ Line up the units in columns. Remember to write the larger quantity on top.

▶ If necessary, regroup the values of the first number using equivalent values.

▶ Subtract each unit.

▶ Simplify your answer if possible.

Find the difference of 3 hr 25 min and 1 hr 35 min.

$$
\begin{array}{r}
3 \text{ hr } 25 \text{ min} \\
-1 \text{ hr } 35 \text{ min}
\end{array}
$$

Since 60 min = 1 hr

3 hr 25 min = 2 hr + 60 min + 25 min

= 2 hr 85 min

$$
\begin{array}{r}
-1 \text{ hr } 35 \text{ min} \quad = \underline{1 \text{ hr } 35 \text{ min}} \\
1 \text{ hr } 50 \text{ min}
\end{array}
$$

LIST 55

Rules for Multiplying Units of Measurement

Multiplying units of measurement is similar to ordinary multiplication. The final step, however, is to simplify your answer.

▶ Set up your problem like ordinary multiplication.

▶ Multiply from right to left, multiplying each column separately.

▶ If necessary, convert the units.

Find the product of 3 and 4 lb 9 oz.

$$\begin{array}{r} 4 \text{ lb } 9 \text{ oz} \\ \times 3 \\ \hline 12 \text{ lb } 27 \text{ oz} \end{array}$$

Since 16 oz = 1 lb

$$\begin{array}{r} 1r11 \\ 16{\overline{)27}} \\ \underline{16} \\ 11 \end{array}$$

27 oz = 1 lb 11 oz

12 lb 27 oz = 12 lb + 1 lb 11 oz = 13 lb 11 oz

LIST 56

Rules for Dividing Units of Measurement

The following steps are helpful for dividing units of measurement.

▶ Convert the quantity into the smallest unit.

▶ Divide.

▶ If possible, express your quotient in terms of the largest quantity.

Divide 9 ft 4 in by 2.

Since 1 ft = 12 in

9 ft = 9 × 12 in = 108 in

9 ft 4 in = 108 in + 4 in = 112 in

$$\begin{array}{r} 56 \text{ in} \\ 2{\overline{)112}} \end{array}$$

$$\begin{array}{r} 4r8 \\ 12{\overline{)56}} \\ \underline{48} \\ 8 \end{array}$$

56 in = 4 ft 8 in

LIST 57

English–Metric Equivalents

Since the United States uses both the English and metric systems of measurement, it is often helpful to know what a particular unit in the English system equals in metrics. (Note that the values for fluid ounces, quarts, and gallons on this list are U.S. units, which vary slightly from British units.)

Length

1 inch = 25.4 millimeters

1 inch = 2.54 centimeters

1 foot = 0.3048 meter

1 yard = 0.9144 meter

1 rod = 5.029 meters

1 furlong = 201.17 meters

1 mile = 1.6093 kilometers

1 nautical mile = 1.852 kilometers

Area

1 square inch = 6.4516 square centimeters

1 square foot = 929.03 square centimeters

1 square foot = 0.092903 square meter

1 square yard = 0.8361 square meter

1 acre = 0.4047 hectare

1 square mile = 258.999 hectares

1 square mile = 2.5899 square kilometers

Volume

1 cubic inch = 16.387 cubic centimeters

1 cubic foot = 0.0283 cubic meter

1 cubic yard = 0.765 cubic meter

LIST 57

(Continued)

Liquid Measure

> 1 teaspoon = 4.9289 milliliters
>
> 1 fluid ounce = 29.573 milliliters
>
> 1 cup = 0.237 liter
>
> 1 pint = 0.4732 liter
>
> 1 quart = 0.9463 liter
>
> 1 gallon = 3.7853 liters

Weight (Avoirdupois)

> 1 grain = 64.7989 milligrams
>
> 1 dram = 1.772 grams
>
> 1 ounce = 28.350 grams
>
> 1 pound = 453.59237 grams
>
> 1 pound = 0.45359 kilogram
>
> 1 short ton = 0.907 metric ton

LIST 58

Metric–English Equivalents

The following list shows what common metric units equal in the English system.

Length

1 millimeter = 0.03937 inch

1 centimeter = 0.3937 inch

1 decimeter = 3.937 inches

1 meter = 39.37 inches

1 meter = 3.28 feet

1 meter = 1.094 yards

1 kilometer = 3,280.8 feet

1 kilometer = 0.621 mile

Area

1 square centimeter = 0.15499 square inch

1 square meter = 10.764 square feet

1 square meter = 1.196 square yards

1 hectare = 2.471 acres

1 square kilometer = 0.386 square mile

Volume

1 cubic centimeter = 0.06102 cubic inch

1 cubic meter = 61,023.744 cubic inches

1 cubic meter = 35.315 cubic feet

1 cubic meter = 1.308 cubic yards

Liquid Measure

1 milliliter = 0.033814 fluid ounce

1 centiliter = 0.338 fluid ounce

1 liter = 1.0567 quarts

1 kiloliter = 264.172 gallons

Weight

1 milligram = 0.00003527 ounce (avoirdupois)

1 gram = 0.035 ounce (avoirdupois)

1 kilogram = 2.204623 pounds (avoirdupois)

1 metric ton = 2,204.623 pounds (avoirdupois)

1 metric ton = 1.1023 short tons

LIST 59

Conversion Factors for Length

Because we use both the English system and the metric system for measuring length, it is helpful to be able to convert measurements from one system to the other.

Original Measurement	Converted to	Multiply by
inches	millimeters	25.4
inches	centimeters	2.54
inches	meters	0.0254
feet	meters	0.3048
yards	meters	0.9144
rods	meters	5.0292
furlongs	meters	201.168
miles	kilometers	1.60934
nautical miles	kilometers	1.852
millimeters	inches	0.03937
centimeters	inches	0.3937
meters	inches	39.3701
meters	feet	3.2808
meters	yards	1.09361
kilometers	miles	0.621371
kilometers	nautical miles	0.539957

LIST 60

Conversion Factors for Area

The following conversion factors are useful for converting area in the English and metric systems.

Original Measurement	Converted to	Multiply by
square inches	square centimeters	6.4516
square yards	square meters	0.836127
square rods	square meters	25.293
square miles	square kilometers	2.58999
acres	square meters	4,046.86
square centimeters	square inches	0.15499
square meters	square feet	10.7639
square meters	square yards	1.19599
square kilometers	square miles	0.386019

LIST 61

Conversion Factors for Volume

When converting units of volume in the English and metric systems, use the following factors.

Original Measurement	Converted to	Multiply by
cubic inches	cubic centimeters	16.387
cubic feet	cubic meters	0.0283168
cubic feet	liters	28.3168
cubic yards	cubic meters	0.764555
cubic centimeters	cubic inches	0.06102
cubic meters	cubic feet	35.314
cubic meters	cubic yards	1.308

LIST **62**

Conversion Factors for Liquid Capacity

Use the following factors for converting the English and metric systems for liquid capacity.

Original Measurement	Converted to	Multiply by
teaspoons	milliliters	4.929
tablespoons	milliliters	14.787
gills	milliliters	118.294
fluid ounces	milliliters	29.573
pints	liters	0.4732
quarts	liters	0.9463
gallons	liters	3.7853
milliliters	teaspoons	0.2029
milliliters	tablespoons	0.0676
milliliters	fluid ounces	0.0338
centiliters	fluid ounces	0.338
liters	quarts	1.0567
liters	gallons	0.26418
kiloliters	gallons	264.18

LIST **63**

Conversion Factors for Dry Capacity

The following factors are handy for converting English and metric units of dry capacity.

Original Measurement	Converted to	Multiply by
pints	liters	0.5505
quarts	liters	1.1012
pecks	liters	8.809
bushels	liters	35.239
liters	quarts	0.9081
dekaliters	pecks	1.1351
hectoliters	bushels	2.8378
kiloliters	bushels	28.378

LIST 64

Conversion Factors for Mass (Weight)

Use the following factors for converting the English and metric systems for mass (weight).

Original Measurement	Converted to	Multiply by
carats	milligrams	200.0
grains	milligrams	64.799
ounces (avoirdupois)	grams	28.3495
pounds (avoirdupois)	grams	453.59237
pounds (avoirdupois)	kilograms	0.453592
tons (short, 2,000 lbs)	kilograms	907.185
tons (short)	metric ton	0.907
tons (long, 2,240 lbs)	kilograms	1,016.047
tons (long)	metric ton	1.016
milligrams	grains	0.015
grams	ounces (avoirdupois)	0.035
kilograms	pounds (avoirdupois)	2.204623
metric ton	short ton	1.102
metric ton	long ton	0.984

LIST 65

Measures of Force and Pressure

Force and pressure are closely related. *Force* is anything that changes the motion or state of rest in a body. *Pressure* is a force acting on a surface per unit of area.

Dyne—the force needed to accelerate a 1-gram mass 1 centimeter per second squared.

   1 dyne = 0.0000723 poundal

   1 dyne = 10^{-5} newtons

Poundal—the force needed to accelerate a 1-pound mass 1 foot per second squared.

   1 poundal = 13,825.5 dynes

   1 poundal = 0.138255 newton

Newton—the force needed to accelerate a 1-kilogram mass 1 meter per second squared.

   1 newton = 10^5 dynes

   1 newton = 7.23300 poundals

Pascal—unit used to measure pressure. One pascal equals 1 newton per square meter or 0.020855 pound per square foot.

Atmosphere—unit used to measure pressure at sea level. One atmosphere equals 14.6952 pounds per square inch; 2,116.102 pounds per square foot; 1.0332 kilograms per square centimeter; 101,325 newtons per square meter.

LIST 66

Measurement of Time

Time. It's so important to us that we have created numerous ways to measure it.

nanosecond = 1 billionth of a second (sec)

microsecond = 1 millionth of a second

millisecond = 1 thousandth of a second

60 seconds = 1 minute (min)

60 minutes = 1 hour (hr)

24 hours = 1 day (d)

7 days = 1 week (wk)

4 weeks = approximately 1 month (mn)

52 weeks = 1 year (yr)

30 days = 1 calendar month (individual months vary)

12 months = 1 year

365 days = 1 common year*

366 days = 1 leap year

10 years = 1 decade

100 years = 1 century

1,000 years = a millennium

Here are some other "timely" terms:

Sidereal Time—Also known as astronomical time, this method of measurement uses the movement of the stars to calculate time. An average sidereal day is 23 hours, 56 minutes, and 4.09 seconds long.

Solar Time—An outmoded measurement of time in which noon occurs when the sun is at its highest point over a given location. Of course, noon (and therefore time) varies from place to place.

Atomic Time—Atomic clocks calculate time with extreme accuracy based on the natural resonance frequency of the cesium-133 atom.

Standard Time—Introduced in 1883 by international agreement, Standard Time divided the Earth into 24 time zones. Calculated on solar time, the base of the system is the zero meridian that passes through the Royal Greenwich Observatory at Greenwich, England. Time is measured east or west of this Prime Meridian according to the time zones.

Daylight Saving Time—In the United States, Standard Time plus one hour. Daylight Saving Time seems to make the day last longer by an hour.

---

*More precisely, a year equals $365\frac{1}{4}$ days. To keep the calendar accurate, an extra day is added every four years, resulting in a leap year.

LIST **67**

Temperature Formulas

Three scales are used for measuring temperature. The most common are the Celsius and Fahrenheit scales. The third, the Kelvin scale, is often used to measure the temperature in scientific experiments.

It is sometimes necessary to convert from one scale to the other. Below are four formulas that may be used. Note that for the conversions to Celsius and Fahrenheit, the formulas are written in both fraction and decimal forms.

To convert from degrees Celsius (°C) to degrees Fahrenheit (°F) use:

$$F = \frac{9C}{5} + 32$$

$$F = (C \times 1.8) + 32$$

To convert from degrees Fahrenheit (°F) to degrees Celsius (°C) use:

$$C = \frac{F - 32}{9} \times 5$$

$$C = (F - 32) \div 1.8$$

To convert from degrees Kelvin (°K) to degrees Celsius (°C) use:

$$C = K - 273.15$$

To convert from degrees Celsius (°C) to degrees Kelvin (°K) use:

$$K = C + 273.15$$

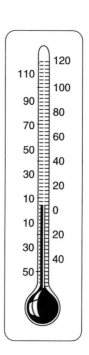

Following is a short list of some equivalent temperatures:

water boils	100°C	=	212°F
normal body temperature	37°C	=	98.6°F
water freezes	0°C	=	32°F
absolute zero	− 273.15°C	=	− 459.67°F = 0°K

LIST 68

Getting a Feel for Temperature

The two most common scales used for measuring temperature are the Celsius and Fahrenheit scales. On the Celsius scale, named after Anders Celsius (1701–1744), the freezing point of water is 0° and the boiling point of water is 100°. On the Fahrenheit scale, named after Gabriel Fahrenheit (1686–1736), the freezing point of water is 32° and the boiling point is 212°.

Because the scales denote the freezing and boiling points of water at different degrees, people who are familiar with one scale usually have trouble understanding the other. Following are some degrees that you can use as reference points in comparing Celsius temperatures to Fahrenheit temperatures. Note that in some cases, the conversions are approximations.

Degrees Celsius	Degrees Fahrenheit	Reference Point
327°C	621°F	lead melts
190°C	374°F	a hot oven
100°C	212°F	water boils
80°C	176°F	hot soup
66°C	151°F	hot faucet water
60°C	140°F	broiled steak
58°C	137°F	highest air temperature recorded on Earth
45°C	113°F	a hot bath
40°C	104°F	a high fever
38°C	101°F	a warm bath
37°C	98.6°F	normal body temperature
35°C	95°F	a hot day
20°C	68°F	room temperature on a winter day
10°C	50°F	a cool fall day
7°C	45°F	cold water
1°C	33°F	very cold water, just above freezing
0°C	32°F	water freezes
–5°C	23°F	a snowy day
–11°C	12°F	frozen yogurt

LIST **69**

The Richter Scale

The Richter Scale, developed by seismologist Charles Richter, is used to express the amount of energy released at the focus of an earthquake. The scale is logarithmic and based on a numerical system of exponents. For example, the difference between 6.0 and 7.0 on the Richter Scale is not one but a factor of ten. Thus, an earthquake that measures 6.0 on the Richter Scale is a hundred times more powerful than an earthquake that measures 4.0. A quake of 8.0 is 10 million times greater than a 1.0 quake.

Richter Number	*Magnitude Increase*	*Comment*
9	100,000,000	near total devastation
8	10,000,000	a disaster, few buildings left standing
7	1,000,000	many buildings destroyed
6	100,000	buildings shake; roads and walls crack
5	10,000	strong rumbling; china and dishes break
4	1,000	weak; much like a passing truck
3	100	very weak; less than 3.5
2	10	detectable only by seismographs
1	1	

The great San Francisco earthquake of 1906 had a magnitude of 8.3. It destroyed the city. This was hardly a burp compared to the earthquake that accompanied the volcanic explosion and sinking of the island of Krakatoa in 1883, which some experts estimate would have measured 9.9 on the Richter Scale.

Wind Speed

Description of the wind can be rather inaccurate, depending on an individual's point of view. To give some order to describing the wind, Admiral Sir Francis Beaufort devised the following wind speed scale in 1805. Named the Beaufort Scale, it is still used today.

	Wind Speed		
Description of Air	*MPH*	*Knots*	*Km/hr*
0. Calm	below 1	below 1	below 1
1. Light Air	1–3	1–3	1–5
2. Light Breeze	4–7	4–6	6–11
3. Gentle Breeze	8–12	7–10	12–19
4. Moderate Breeze	13–18	11–16	20–28
5. Fresh Breeze	19–24	17–21	29–38
6. Strong Breeze	25–31	22–27	39–49
7. Near Gale	32–38	28–33	50–61
8. Gale	39–46	34–40	62–74
9. Strong Gale	47–54	41–47	75–88
10. Storm	55–63	48–55	89–102
11. Violent Storm	64–73	56–63	103–117
12. Hurricane	74+	64+	118+

LIST 71

The Saffir-Simpson Hurricane Scale

The Saffir-Simpson Hurricane Scale offers a description of a hurricane's intensity. The scale ranges from 1 to 5, with a Category 1 hurricane being the weakest and a Category 5 being the strongest.

Category 1

Wind Speed: 74–95 mph, 119–153 km/hr, 64–82 knots

Comment: Most buildings will experience minimal or no damage. There will be some damage to mobile homes, shrubbery, and trees. A storm surge of 4 to 5 feet above normal will result in some coastal flooding, generally minor.

Category 2

Wind Speed: 96–110 mph, 154–177 km/hr, 83–95 knots

Comment: Some roof and window damage can be expected. Some trees will be uprooted. Some mobile homes and piers will sustain significant damage. A storm surge of 6 to 8 feet will result in flooding of low-lying areas.

Category 3

Wind Speed: 111–130 mph, 178–209 km/hr, 96–113 knots

Comment: Large trees will be blown down, most mobile homes will be destroyed, and many buildings will suffer roof and window damage. A storm surge of 9 to 12 feet may cause major flooding.

Category 4

Wind Speed: 131–155 mph, 210–249 km/hr, 114–135 knots

Comment: Mobile homes will be completely destroyed, and many buildings will suffer significant damage. Many trees will be blown down. A storm surge of 13 to 18 feet will likely cause extensive flooding and major damage to the lower floors of buildings near the shore. Massive evacuation may be required for residential coastal areas, in some cases up to 6 miles inland.

Category 5

Wind Speed: greater than 155 mph, 249 km/hr, 135 knots

Comment: Massive widespread damage should be expected. Many buildings will suffer complete roof and structural failures. A storm surge of over 18 feet will result in major damage to the lower floors of all structures along the coastline that are less than 15 feet above sea level. Extensive flooding will likely require evacuation of residential areas on low ground up to 10 miles inland.

LIST **72**

The Fujita Scale of Tornado Intensity

The Fujita Scale provides a description of the likely damage caused by tornadoes of varying intensities. The scale ranges from 0 to 5, with an F5 tornado being the most destructive.

F0

Wind Speed: 40–72 mph, 64–116 km/hr

Comment: Light damage. Some branches may be broken, trees with shallow roots may be blown over, and chimneys and billboards may be damaged.

F1

Wind Speed: 73–112 mph, 117–181 km/hr

Comment: Moderate damage. Some roofs and windows may suffer damage, mobile homes may be pushed off their foundations, and trees may be blown over.

F2

Wind Speed: 113–157 mph, 182–253 km/hr

Comment: Considerable damage. There will be much structural damage to buildings; mobile homes will be destroyed. Large trees may be uprooted or snapped.

F3

Wind Speed: 158–206 mph, 254–332 km/hr

Comment: Severe damage can be expected. Houses suffer significant damage. Trains may be overturned, and automobiles may be lifted off the ground and thrown through the air. Many trees will be uprooted.

F4

Wind Speed: 207–260 mph, 333–418 km/hr

Comment: Widespread destruction. Most houses will be destroyed. Large objects will be lifted from the ground and thrown about.

F5

Wind Speed: 261–318 mph, 419–512 km/hr

Comment: Massive devastation. Houses will be lifted off their foundations and destroyed. Automobiles may be thrown up to 300 feet through the air. Trees that are left standing will be debarked.

LIST **73**

Computer Memory

Computer memory is based on bits and bytes.

8 bits = 1 character = 1 byte

Since one character might correspond to the letter *a,* 1 megabyte of memory has the capacity of storing 1 million single letters. Not too long ago, a megabyte was considered to be a lot of memory, but not any more. Following is a rundown of computer memory.

1 kilobyte = 1,024 bytes or 2^{10}

1 megabyte = 1,048,576 bytes or 2^{20}

1 gigabyte = 1,073,741,824 bytes or 2^{30}

1 terabyte = 1,099,511,627,776 or 2^{40}

1 petabyte = 1,125,899,906,842,624 bytes or 2^{50}

Terabyte databases are becoming common, and petabyte databases are on the horizon, with even greater storage capacity to eventually follow.

LIST 74

Paper Measures or Paper Weights

Upon hearing the phrase "paper weights," many people think of the decorative objects that are used to keep papers from flying off desks when a breeze gusts in through a window or someone shuts a door too fast. But paper weight also refers to the thickness of paper. Sheets might be 20-pound paper, 60-pound paper, or more.

The way paper is packaged can vary too. Most people have heard that a ream of paper contains 500 sheets. It can; but it can also contain 480.

The following list clarifies the confusion.

Some Standard Amounts

24 sheets = 1 quire

25 sheets = 1 printer's quire

20 quires = 1 ream

21.5 quires = 1 printer's ream

2 reams = 1 bundle

4 bundles = 1 case

4 reams = 1 printer's bundle

10 reams = 1 bale

480 sheets = 1 short ream

500 sheets = 1 long ream

Guidelines for Commercial Paper Weights

9 pounds—onionskin paper

20 pounds—standard copier paper or typing paper

24 pounds—standard letterhead paper

60 pounds—thick enough for printing on both sides because the ink will not bleed through

65 pounds—typical business cards

120 pounds—poster paper

U.S. Money: Coins and Bills

The following coins and bills are in current circulation in the United States.

$ .01	1¢	penny	$\frac{1}{100}$ of a dollar
$ .05	5¢	nickel	$\frac{1}{20}$ of a dollar
$ .10	10¢	dime	$\frac{1}{10}$ of a dollar
$ .25	25¢	quarter	$\frac{1}{4}$ of a dollar
$ .50	50¢	half dollar	$\frac{1}{2}$ of a dollar
$ 1.00		coin	
$ 1.00		dollar	
$ 2.00		two dollars*	
$ 5.00		five dollars	
$ 10.00		ten dollars	
$ 20.00		twenty dollars	
$ 50.00		fifty dollars	
$ 100.00		one hundred dollars**	

*The $2.00 bill is seldom used.
**Bills bigger than $100.00 are no longer issued.

Section Three

Geometry

LIST 76

Undefined Terms of Geometry

Geometry is the mathematics of properties, measurement, and the relationships of points, lines, angles, surfaces, and solids in space. It is based on three undefined terms: *point, line,* and *surface.* An understanding of these terms is necessary to understand other geometric terms and principles. Although *point, line,* and *surface* remain undefined, they can be described.

Point

▶ Is represented by a dot.

▶ Has no length, width, or thickness.

▶ Is designated by a capital letter next to the dot.

Point A

Line

▶ Has only length.

▶ Has no width or thickness.

▶ Is unlimited and extends infinitely in either direction.

▶ Is designated by two capital letters, which represent two points on the line.

▶ May also be designated by a small letter.

▶ Can be classified as:

    straight

    curved

    broken

    combination

▶ Unless stated otherwise, is understood to be straight.

▶ Is the shortest distance between two points.

Line $\overleftrightarrow{AB}$

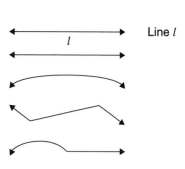

Line l

Surface

▶ Has length and width.

▶ Does not have thickness.

▶ May be represented by the side of a box, outside of a sphere, a wall with no windows or other openings.

▶ Is a "plane" surface when it contains a straight line that connects any two points on the plane. A plane surface is sometimes denoted by a closed, four-sided figure with a capital letter at its vertex.

Plane K

Copyright © 2005 by Judith A. Muschla and Gary Robert Muschla

LIST 77

Euclid's Axioms and Postulates

In his most famous work, *Elements,* Euclid, a Greek mathematician who lived around 300 B.C., described geometry as a formal system of reasoning. Formal geometry is based on *axioms,* a set of "common notions," which are accepted to be true without proof. They apply to mathematics in general.

Postulates are assumptions that apply to a specific branch of mathematics—in this case, geometry. Using axioms and postulates, a person can define new terms and prove statements about them using deductive reasoning. These statements are called *theorems.*

Euclid's Axioms

1. Things equal to the same thing are equal to each other.
2. If equals are added to equals, the sums are equal.
3. If equals are subtracted from equals, the differences are equal.
4. Things that coincide with one another are equal.
5. The whole is greater than any of its parts.

Euclid's Postulates

1. A straight line can be drawn between any two points.
2. Any straight line segment can be extended infinitely.
3. A circle with any radius may be described around a point as a center.
4. All right angles are equal to each other.
5. If a straight line falling on two straight lines makes the interior angles on the same side together less than two right angles, the two straight lines, if produced indefinitely, meet on that side on which the angles are together less than two right angles. (Because this fifth postulate is not felt to be a common notion and fit the definition of a postulate, John Playfair, a Scottish mathematician and physicist in the early nineteenth century, offered a revision of it: *Given a line and a point not on the line, there is one and only one line parallel to the given line.* Playfair's postulate is often substituted for the fifth postulate of Euclid.)

LIST 78

Lines and Planes

A *line* is one of the undefined terms in geometry. A description of a line is that it has length but no thickness or depth. In theory, a line may be extended infinitely in each direction.

A *plane* is a flat surface that extends infinitely in all directions. Imagine extending the length and width of a table top forever.

Lines that lie in the same plane are called *coplanar lines*. Any two coplanar lines must have one and only one of the characteristics listed below.

▶ The lines may intersect. If they intersect and form right angles, they are perpendicular lines.

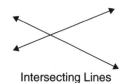
Intersecting Lines

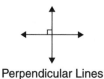
Perpendicular Lines

▶ The lines may be parallel. Parallel lines will never meet.

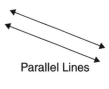
Parallel Lines

▶ The lines may coincide. Lines that coincide are actually the same line.

Lines That Coincide

Lines that lie in different planes and do not intersect are called *noncoplanar lines* or *skew lines*.

▶ If two planes do not intersect, the planes are parallel.

Skew Lines

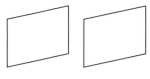
Parallel Planes

▶ If two planes intersect, their intersection is a line.

Intersecting Planes

Steps for Naming an Angle

An *angle* is formed by two rays that have the same *endpoint,* which is called the *vertex* of the angle. The rays are called the sides of the angle. (A *ray* is a part of a line consisting of a given point called the endpoint. The ray continues forever in the other direction.) A point between the sides of the angle is in the interior of the angle. ∠ is the symbol for angle.

There are three ways to name an angle.

1. Use three letters. The center letter corresponds to the vertex. The other two letters are points on each ray. The angle shown here can be named ∠ABC or ∠CBA. It can be read as "angle ABC" or "angle CBA."

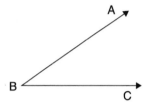

2. Place a number or letter at the vertex in the interior of the angle. The angle shown here can be named ∠1. It is read as "angle 1."

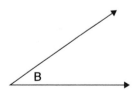

3. Use a single capital letter at the vertex in the interior of the angle. This may be used only if there is one angle at the vertex. The angle shown here can be named ∠B and read as "angle B."

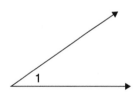

LIST **80**

Measuring Angles

The size of an angle has nothing to do with the lengths of its sides, which technically extend forever. The size of an angle depends on the "opening" of the angle.

An angle is measured in *degrees* with an instrument called a *protractor.*

To measure an angle:

1. Compare its side to the corner of this page. The corner represents a *right angle,* whose measurement is 90°.

   ▶ If the opening of the angle is smaller than the corner, the angle is an *acute angle.* Its measure is less than 90°.

   ▶ If the opening is larger than the corner of the page, the angle is an *obtuse* angle. Its measure is more than 90°.

2. Locate the point of your protractor, and place it over the vertex of the angle. (The *vertex* is the point from which the two rays that make up the sides of the angle begin.)

3. Rotate the protractor, keeping the vertex aligned until one side of the angle is on the 0°–180° line of the protractor.

4. Read the angle measure that is determined by the side of the angle that is not on the 0°–180° line of the protractor. You may have to extend one side of the angle so that it crosses the scale.

   ▶ If the angle is acute, read the smaller scale.

   ▶ If the angle is obtuse, read the larger scale.

   ▶ If the angle is a right angle, the measure is 90°.

5. Use the proper notation.

   ▶ *m* is the symbol for "measure of." *Example: m∠ABC* = 42° is read "the measure of angle *ABC* is 42°."

Drawing Angles

To draw an angle, you will need a protractor and a ruler. Follow these steps:

1. Use your ruler (or the straight edge of your protractor) to draw a ray on your paper.

2. Align the point on the protractor with the end of the ray, which will become the vertex of the angle.

3. Using the right side of the protractor, align the ray with the 0°–180° line on the protractor.

4. Use the inner scale to find the measure of the angle you wish to draw. If the endpoint of the ray is located to the right of the ray (←), use the outer scale.

5. Mark that point on your paper.

6. Using your ruler, draw a ray from the vertex to the point you marked.

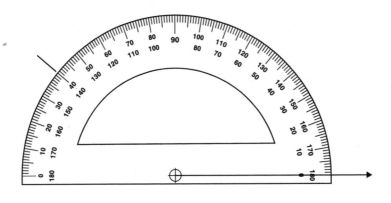

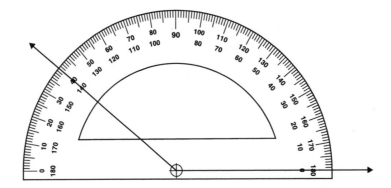

LIST 82

Types of Angles and the Facts

Five types of angles are essential to the study of geometry. Once you understand these angles, you can build on your knowledge and skills.

Kinds of Angles

▶ *Acute angle*—an angle whose measure is less than 90°.

▶ *Right angle*—an angle whose measure equals 90°. A box in the vertex denotes a right angle.

▶ *Obtuse angle*—an angle whose measure is greater than 90° and less than 180°.

▶ *Straight angle*—an angle whose measure equals 180°.

▶ *Reflex angle*—an angle whose measure is greater than 180° and less than 360°.

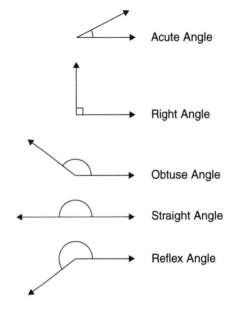

Acute Angle

Right Angle

Obtuse Angle

Straight Angle

Reflex Angle

Angle Facts

▶ Equal angles are angles that have the same number of degrees.

▶ A ray that bisects an angle divides it into two equal parts. The line is called the *angle bisector.*

▶ Congruent angles have the same measure.

▶ Perpendiculars are lines that form right angles.

▶ All right angles are congruent.

▶ The sides of a straight angle lie on a straight line.

▶ All straight angles are congruent.

▶ A perpendicular bisector of a line bisects the line and is perpendicular to the line.

LIST 83

Facts About Pairs of Angles

A group of two angles is known as a pair of angles. Special angle pairs have their own names and distinguishing features.

▶ *Adjacent angles*—two angles that have the same vertex and a common side.

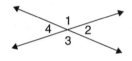

∠ABC and ∠CBD are Adjacent Angles.

▶ *Vertical angles*—two nonadjacent angles formed by two intersecting lines.

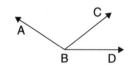

∠2 and ∠4 are Vertical Angles. ∠1 and ∠3 are Vertical Angles.

▶ *Complementary angles*—two angles whose measures add up to 90°. One angle is the complement of the other.

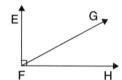

∠EFG and ∠GFH are Complementary Angles.

▶ *Supplementary angles*—two angles whose measures add up to 180°. One angle is the supplement of the other.

There are 4 pairs of Supplementary Angles in the second diagram:

∠1 and ∠2   ∠1 and ∠4
∠2 and ∠3   ∠4 and ∠3

Here are some additional facts about pairs of angles:

▶ If an angle is cut into two adjacent angles, then the sum of the measures of the adjacent angles equals the measure of the original angle.

▶ If the exterior sides of a pair of adjacent angles are perpendicular, then the angles are complementary.

▶ If the exterior sides of a pair of adjacent angles form a straight line, then the angles are supplementary.

▶ If two angles are congruent and supplementary, then each angle is a right angle.

▶ Complements of the same or congruent angles are congruent.

▶ Supplements of the same or congruent angles are congruent.

▶ Vertical angles are congruent.

LIST 84

Angles Formed by a Transversal

A *transversal* is a line that cuts across two or more lines in the same plane. As it does, it forms five types of angles.

1. *Interior angles*—angles inside the region bounded by the two lines.

   ▶ ∠1, ∠2, ∠3, and ∠4 are interior angles.

2. *Exterior angles*—angles outside the region bounded by the two lines.

   ▶ ∠5, ∠6, ∠7, and ∠8 are exterior angles.

3. *Alternate interior angles*—two nonadjacent interior angles that are on opposite sides of the transversal.

   ▶ ∠1 and ∠4 are alternate interior angles.
   ▶ ∠2 and ∠3 are alternate interior angles.

4. *Alternate exterior angles*—two nonadjacent exterior angles that are on opposite sides of the transversal.

   ▶ ∠5 and ∠8 are alternate exterior angles.
   ▶ ∠6 and ∠7 are alternate exterior angles.

5. *Corresponding angles*—two nonadjacent angles (one is an interior angle and the other is an exterior angle) on the same side of the transversal.

   ▶ Corresponding angles are
      ∠3 and ∠5
      ∠1 and ∠7
      ∠6 and ∠4
      ∠2 and ∠8

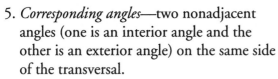

LIST 85

Principles of Parallel Lines

Parallel lines are lines that are in the same plane and never intersect. Imagine the lines made by the rails of a straight train track that extend forever. The distance between the lines remains constant.

$\parallel$ is the symbol for "is parallel to."

$\nparallel$ is the symbol for "is not parallel to."

$l_1 \parallel l_2$ is read "line one is parallel to line two."

Below are some properties of parallel lines:

The Parallel Postulate: Through a given point not on a line, exactly one line may be drawn parallel to the line.

If two lines are parallel, then:

- ▶ Corresponding angles are congruent.
- ▶ Alternate interior angles are congruent.
- ▶ Alternate exterior angles are congruent.
- ▶ Each pair of interior angles on the same side of the transversal is supplementary.
- ▶ A line perpendicular to one of the parallel lines is perpendicular to the other.
- ▶ A line parallel to one of the parallel lines is parallel to the other.

To prove lines are parallel, show that one of the following is true:

- ▶ A pair of corresponding angles is congruent.
- ▶ A pair of alternate interior angles is congruent.
- ▶ A pair of alternate exterior angles is congruent.
- ▶ A pair of interior angles on the same side of a transversal is supplementary.
- ▶ The lines are perpendicular to the same line.
- ▶ The lines are parallel to the same line.

LIST 86

Principles of Perpendicular Lines

Perpendicular lines intersect to form right angles. A line that is perpendicular to a segment and intersects the segment at its midpoint is called the *perpendicular bisector* of the segment.

⊥ is the symbol for perpendicular.

The following are principles of perpendicular lines:

▶ The lines intersect to form right angles.
▶ There is only one perpendicular to a given line through a given point on the line.
▶ There is only one perpendicular to a given line through a given point not on the line.
▶ The distance between a line and a point not on the line is the length of the perpendicular segment drawn from the point to the line.

To prove that two lines are perpendicular, show at least one of the following:

▶ The lines intersect to form a right angle.
▶ The lines intersect to form a pair of congruent adjacent angles.

LIST 87

Types of Polygons

Polygons are closed, plane figures bounded by line segments. The following list contains useful information about polygons. *n* stands for the number of sides.

Polygon	Number of Sides	Number of Diagonals	Sum of Interior Angles*
Triangle	3	0	180°
Quadrilateral	4	2	360°
Pentagon	5	5	540°
Hexagon	6	9	720°
Heptagon	7	14	900°
Octagon	8	20	1,080°
Nonagon	9	27	1,260°
Decagon	10	35	1,440°
Undecagon	11	44	1,620°
Dodecagon	12	54	1,800°
N-gon	n	$\left(n^2 - 3n\right) \div 2$	$(n-2)\left(180°\right)$

*The sum of the exterior angles for each polygon is 360°.

Regular Polygons: Terms and Formulas

A *regular polygon* is equilateral (all sides are congruent) and equiangular (all angles are congruent). While some terms and formulas apply to all polygons, others apply only to regular polygons.

Terms

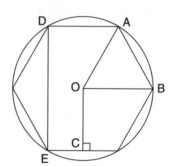

▶ *Center of a regular polygon*—the common center of the inscribed and circumscribed circles. O is the center.

▶ *Radius of a regular polygon*—the line segment that joins the center of the polygon to any vertex. $\overline{OA}$ is a radius.

▶ *Central angle of a regular polygon*—an angle included between two radii drawn to successive vertices. $\angle AOB$ is a central angle.

▶ *Apothem of a regular polygon*—a line segment from the center of the polygon that is perpendicular to one of its sides. $\overline{OC}$ is an apothem.

▶ *Diagonal of any polygon*—a line segment drawn from one vertex to another nonadjacent vertex. $\overline{DE}$ is a diagonal.

Formulas

1. The measure of each interior angle of a regular polygon is $\dfrac{180°\,(n-2)}{n}$, where n is the number of sides.

2. The measure of each exterior angle of a regular polygon is $\dfrac{360°}{n}$, where n is the number of sides.

3. The measure of each central angle of a regular polygon is $\dfrac{360°}{n}$, where n is the number of sides.

4. The sum of the interior angles of any polygon equals $(n-2)\,(180°)$, where n is the number of sides.

5. The sum of the exterior angles of any polygon is 360°.

6. The number of diagonals of any polygon is $\dfrac{n^2-3n}{2}$, where n is the number of sides.

LIST **89**

Principles of Regular Polygons

Regular polygons share some unique characteristics, as the list below shows.

- ▶ A circle may be circumscribed about any regular polygon.
- ▶ A circle may be inscribed in any regular polygon.
- ▶ An equilateral polygon inscribed in a circle is a regular polygon.
- ▶ The radius of a regular polygon is a radius of the circumscribed circle.
- ▶ The radius of a regular polygon bisects the angle to which it is drawn.
- ▶ Apothems of regular polygons are equal.
- ▶ An apothem of a regular polygon bisects the side to which it is drawn.
- ▶ Regular polygons that have the same number of sides are similar.
- ▶ Corresponding lines of regular polygons that have the same number of sides are in proportion.

Shapes of Numbers

The ancient Greeks, in the spirit of amusement and exploration, sometimes used geometric figures to represent numbers. Whole numbers that can be depicted by geometric figures are called *figurate numbers.* Examples of figurate numbers are listed below.

▶ *Triangular numbers*—numbers that can be represented by a triangular array of dots. (Triangles have three sides.)

 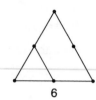

1   3   6

▶ *Square numbers*—numbers that can be represented by a square array of dots. (Squares have four congruent sides.)

 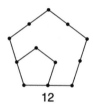

1   4   9

▶ *Pentagonal numbers*—numbers that can be represented by a pentagonal array of dots. (Pentagons have five sides.)

 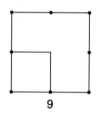

1   5   12

▶ *Hexagonal numbers*—numbers that can be represented by a hexagonal array of dots. (Hexagons have six sides.)

 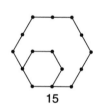

1   6   15

▶ *Heptagonal numbers*—numbers that can be represented by a heptagonal array of dots. (Heptagons have seven sides.)

 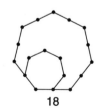

1   7   18

LIST **90**

(Continued)

▶ *Octagonal numbers*—numbers that can be represented by an octagonal array of dots. (Octagons have eight sides.)

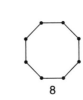 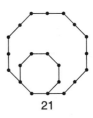

1   8   21

▶ *Nonagonal numbers*—numbers that can be represented by a nonagonal array of dots. (Nonagons have nine sides.)

1   9   24

Some figurate numbers can be represented by arrays that show solid figures. Two of these are:

▶ *Cubic numbers*—numbers that can be represented by a cubic array of dots. (Cubes are solid figures that have six congruent squares as faces.)

 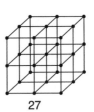

1   8   27

▶ *Tetrahedral numbers*—numbers that can be represented by a tetrahedral array of dots. (Tetrahedra have four congruent triangular faces.)

1   4   10

LIST 91

Figurate Number Formulas

Figurate numbers can be predicted by looking at the array of dots used to represent them. Some patterns are obvious, and others require more study and numerical manipulations. The first five figurate numbers of various shapes are provided in the following list.

Shape	1st	2nd	3rd	4th	5th	nth
Triangular	1	3	6	10	15	$\dfrac{n^2 + n}{2}$
Square	1	4	9	16	25	n^2
Pentagonal	1	5	12	22	35	$\dfrac{3n^2 - n}{2}$
Hexagonal	1	6	15	28	45	$2n^2 - n$
Heptagonal	1	7	18	34	55	$\dfrac{5n^2 - 3n}{2}$
Octagonal	1	8	21	40	65	$3n^2 - 2n$
Nonagonal	1	9	24	46	75	$\dfrac{7n^2 - 5n}{2}$
Cubic	1	8	27	64	125	n^3
Tetrahedral	1	4	10	20	35	$\dfrac{n^3 + 3n^2 + 2n}{6}$

LIST 92

Polyominoes

A *polyomino* is a plane figure made up of congruent squares connected to each other so that each square shares at least one side with one other square. Polyominoes are classified by a name whose prefix denotes the number of squares that are joined together. The first five polyominoes are listed.

Polyomino	Number of Squares	Number of Arrangements	Arrangements
Monomino	1	1	
Domino	2	1	
Tromino	3	2	
Tetromino	4	5	
Pentomino	5	12	

LIST 93

Pentominoes and Hexominoes

Polyominoes are figures formed by joining congruent squares along the sides so that the sides are adjacent. *Pentominoes* are a special type of polyomino that consist of joining five congruent squares. *Hexominoes* are another special polyomino that consist of joining six congruent squares.

Pentominoes and hexominoes can be arranged in a variety of patterns. Some of the most interesting are those that may form boxes. Pentominoes form open boxes; hexominoes form closed boxes or cubes.

Of the 12 pentominoes, 8 can be arranged so that they form an open box when cut out and folded along the dots. These are shown below.

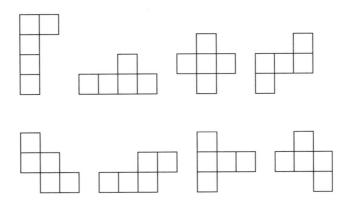

Of the 35 hexominoes, 11 can be arranged to form a closed box or cube when they are cut and folded along the dotted lines. These are shown below.

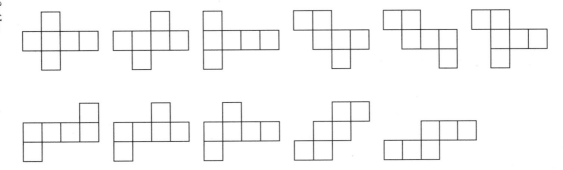

LIST 94

Basic Figures in Geometry

Examples of geometry are all around us. You can find examples of the following geometric figures every day.

Triangles

Triangles are three-sided figures. There are several kinds of triangles.

▶ *Equilateral triangle*—all three sides have the same length, and all three angles have the same measure.

Equilateral Triangle

▶ *Isosceles triangle*—at least two sides have the same length.

Isosceles Triangle

▶ *Scalene triangle*—all three sides have different lengths.

Scalene Triangle

▶ *Acute triangle*—all angles measure less than 90°.

Acute Triangle

▶ *Right triangle*—one angle measures 90°.

Right Triangle

▶ *Obtuse triangle*—one angle measures greater than 90°.

Obtuse Triangle

LIST 94

(Continued)

Quadrilaterals

Quadrilaterals are closed figures with four sides. There are several common quadrilaterals.

▶ *Square*—all sides are the same length; all angles are right angles.

Square

▶ *Rhombus*—all sides are the same length; two pairs of sides are parallel.

Rhombus

▶ *Rectangle*—two pairs of sides are the same length; all angles are right angles.

Rectangle

▶ *Parallelogram*—two pairs of sides have the same length; two pairs of sides are parallel.

Parallelogram

▶ *Trapezoid*—one pair of sides is parallel.

Trapezoid

Circles

▶ A *circle* is a plane figure bounded by a curved line. Every point of the curved line is equidistant from the center.

Circle

Tangram Facts and Figures

A *tangram* is a puzzle that consists of seven geometric shapes. It is pictured below. Facts about tangrams follow.

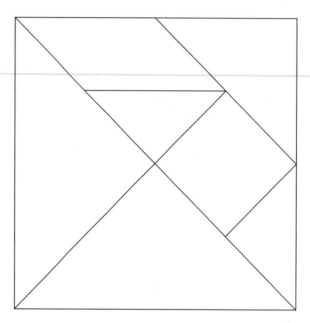

- ▶ The puzzle was invented in China. No one knows exactly when, but it is likely that tangrams have been amusing and puzzling people for hundreds of years.
- ▶ Some people believe that the word *tangram* comes from the word *trangram,* which means "puzzle" or "trinket."
- ▶ Each piece of the puzzle is called a *tan.*
- ▶ A tangram consists of five triangles, a square, and a parallelogram.
- ▶ The two largest triangles are congruent.
- ▶ The two smallest triangles are congruent.
- ▶ All of the triangles are similar.
- ▶ All of the triangles are right isosceles.

LIST 95

(Continued)

▶ The area of each large triangle is $\frac{1}{4}$ the area of the puzzle.

▶ The area of the medium triangle is $\frac{1}{8}$ the area of the puzzle.

▶ The area of each of the small triangles is $\frac{1}{16}$ the area of the puzzle.

▶ The area of the square is $\frac{1}{8}$ the area of the puzzle.

▶ The area of the parallelogram is $\frac{1}{8}$ the area of the puzzle.

▶ The challenge of the puzzle is to reassemble the pieces into the original square.

▶ Thousands of shapes and figures can be formed by arranging the tans. Some are shown below.

LIST **96**

Triangle Terms

The following words are used to describe triangles and their parts.

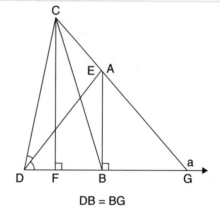

DB = BG

▶ *Triangle*—a closed plane figure formed by three line segments, provided each line segment intersects another line segment at its endpoint. △*CDG* is an example.

▶ *Side*—a line segment joining two endpoints. $\overline{DC}$ is a side.

▶ *Vertex*—the point at which two sides meet. Point *C* is a vertex.

▶ *Vertices*—the plural of vertex.

▶ *Angle of a triangle*—the angle formed inside the triangle by two sides meeting at a vertex. It is also called the interior angle. ∠*DGC* is an interior angle.

▶ *Angle bisector of a triangle*—a line segment that bisects an angle and has one endpoint on the opposite side. $\overline{DE}$ is an angle bisector.

▶ *Exterior angle*—an angle adjacent to an interior angle such that their exterior sides form a straight line. ∠*a* is an exterior angle.

▶ *Perpendicular bisector of a side of a triangle*— a line, ray, or line segment that bisects the side and is perpendicular to it. $\overline{AB}$ is the perpendicular bisector.

▶ *Altitude of a triangle*—a segment drawn from any vertex, perpendicular to the opposite side. It may be extended if necessary. $\overline{CF}$ is an altitude.

▶ *Height*—the length of an altitude.

▶ *Median of a triangle*—a line segment drawn from any vertex of a triangle to the midpoint of the opposite side. $\overline{CB}$ is a median.

LIST 97

Classifying Triangles

Triangles may be grouped according to the measures of their angles and the lengths of their sides.

Triangles Classified According to Angles

1. *Right triangle*—a triangle that has one right angle (an angle equal to 90°).
   ▶ Hypotenuse: the side opposite the right angle.
   ▶ Legs: the two sides adjacent to the right angle.

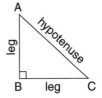

Right Triangle

2. *Obtuse triangle*—a triangle that has one obtuse angle (an angle greater than 90°).

Obtuse Triangle

3. *Acute triangle*—a triangle that has three acute angles (angles less than 90°).

Acute Triangle

4. *Equiangular triangle*—a triangle with three angles whose measures are each 60°.

Equiangular Triangle

Triangles Classified According to Sides

1. *Scalene triangle*—a triangle that has no congruent sides.

Scalene Triangle

2. *Isosceles triangle*—a triangle with at least two congruent sides.
   ▶ Legs: the congruent sides.
   ▶ Base: the remaining side.
   ▶ Base angles: angles on either side of the base and opposite the legs.
   ▶ Vertex: the angle opposite the base.

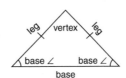

Isosceles Triangle

3. *Equilateral triangle*—a triangle that has three congruent sides.

Equilateral Triangle

LIST **98**

Principles of Triangles

The following principles apply to all triangles.

- ▶ The sum of the measures of the angles of a triangle is 180°.
- ▶ An equilateral triangle has three congruent sides.
- ▶ The measure of each angle in an equiangular triangle is 60°.
- ▶ An equilateral triangle is isosceles and equiangular.
- ▶ A triangle may have at most one right or one obtuse angle.
- ▶ The acute angles of a right triangle are complementary.
- ▶ If two angles of one triangle are congruent to two angles of another triangle, then the remaining pair of angles is congruent.
- ▶ The measure of an exterior angle of a triangle is equal to the sum of the measures of the two nonadjacent interior angles.
- ▶ The measure of an exterior angle of a triangle is greater than the measure of either nonadjacent interior angles.
- ▶ An altitude of a triangle is perpendicular to the side to which it is drawn.
- ▶ Every triangle has three altitudes that intersect at a point called the *orthocenter*. This could happen in one of three ways:
  1. In the interior of an acute triangle
  2. On the right triangle
  3. In the exterior of an obtuse triangle
- ▶ The median of a triangle bisects the side to which it is drawn.
- ▶ Every triangle has three medians, which intersect at a point called the *centroid*.
- ▶ The length of each side of a triangle must be less than the sum of the lengths of the remaining sides.
- ▶ If two sides of a triangle are congruent, then the angles opposite those sides are congruent.
- ▶ If two angles of a triangle are congruent, then the opposite sides are congruent.

Copyright © 2005 by Judith A. Muschla and Gary Robert Muschla

LIST 99

Concurrent Segments of Triangles

Three or more lines (or line segments) that intersect at a common point are said to be *concurrent* at that point. The following line segments are examples of concurrent segments of triangles.

▶ The perpendicular bisectors of the sides are concurrent at a point. This point of concurrency is equidistant from the vertices of the triangle and is called the *circumcenter of the triangle. O* is the circumcenter of the triangle.

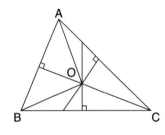

$OA = OB = OC$

▶ Altitudes are concurrent at a point. This point of concurrency is called the *orthocenter of the triangle.* H is the orthocenter of the triangle. The triangle formed by joining the feet of the altitudes is known as the pedal triangle. $\triangle JIK$ is the pedal triangle in the example.

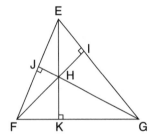

▶ The angle bisectors of a triangle are concurrent at a point that is equidistant from the sides of the triangle. This point of concurrency is the *center of the triangle. R* is the center of the triangle.

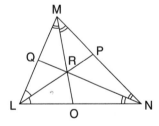

$QR = RP = RO$

▶ The medians of a triangle are concurrent at a point called the *centroid of the triangle.* This point is $\frac{2}{3}$ of the distance from the vertex of the triangle to the midpoint of the side opposite the vertex. *X* is the centroid of the triangle.

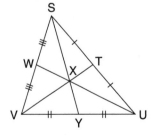

$$VX = \frac{2}{3}VT$$

$$SX = \frac{2}{3}YS$$

$$UX = \frac{2}{3}UW$$

LIST **100**

Special Right Triangles

The 30°–60°–90° triangle is half of an equilateral triangle. The 45°–45°–90° triangle is half of a square. The legs and hypotenuse of these triangles enjoy a special relationship, as shown below.

Relationships in a 30°–60°–90° Triangle

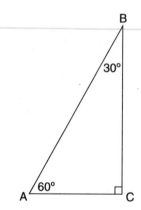

▶ The length of the leg opposite the 30° angle equals $\frac{1}{2}$ the length of the hypotenuse.

$$AC = \frac{1}{2}AB$$

▶ The length of the leg opposite the 60° angle equals $\frac{1}{2}$ the length of the hypotenuse times $\sqrt{3}$.

$$BC = \frac{1}{2}AB\sqrt{3}$$

▶ The length of the leg opposite the 60° angle equals the length of the leg opposite the 30° angle times $\sqrt{3}$.

$$BC = AC\sqrt{3}$$

Relationships in a 45°–45°–90° Triangle

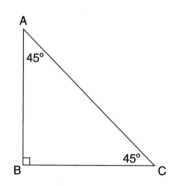

▶ The legs are congruent.

$$AB = BC$$

▶ The length of the hypotenuse equals the length of either leg times $\sqrt{2}$.

$$AC = AB\sqrt{2}$$
$$AC = BC\sqrt{2}$$

▶ The length of either leg is equal to the product of the hypotenuse and $\frac{1}{2}$ and $\sqrt{2}$.

$$AB = \frac{1}{2}AC\sqrt{2}$$
$$BC = \frac{1}{2}AC\sqrt{2}$$

LIST 101

Pythagorean Triples

In the sixth century B.C., Pythagoras discovered a relationship between the hypotenuse of a right triangle and the two legs. $a^2 + b^2 = c^2$ is called the Pythagorean theorem. a and b represent the legs of the triangle, and c represents the hypotenuse.

The most common right triangle is known as the 3–4–5 triangle. This set is called the *primitive set* because the numbers are relatively prime. Multiplying each value by a natural number, for example, 2, 3, or 4, results in another Pythagorean triple such as 6–8–10 or 9–12–15 or 12–16–20.

There are 50 sets of triples where each leg and hypotenuse is less than 100. Sixteen of these are primitive sets and denoted by * in the following list.

Fifty Pythagorean Triples

3–4–5*	5–12–13*	9–40–41*	33–56–65*
6–8–10	10–24–26	18–80–82	
9–12–15	15–36–39		36–77–85*
12–16–20	20–48–52	11–60–61*	
15–20–25	25–60–65		39–80–89*
18–24–30	30–72–78	12–35–37*	
21–28–35	35–84–91	24–70–74	48–55–73*
24–32–40			
27–36–45	8–15–17*	13–84–85*	65–72–97*
30–40–50	16–30–34		
33–44–55	24–45–51	16–63–65*	
36–48–60	32–60–68		
39–52–65	40–75–85	20–21–29*	
42–56–70		40–42–58	
45–60–75	7–24–25*	60–63–87	
48–64–80	14–48–50		
51–68–85	21–72–75	28–45–53*	
54–72–90			
57–76–95			

LIST 102

Proving Triangles Congruent

Congruent triangles have the same size and shape. Two triangles can be proven to be congruent if they agree with any of the statements below. Note that ≅ is the symbol for "is congruent to."

Side-Side-Side (S-S-S)—If three sides of one triangle are congruent to corresponding sides of the other triangle, the two triangles are congruent.

Side-Angle-Side (S-A-S)—If two sides and the included angle of one triangle are congruent to the corresponding parts of the other triangle, the two triangles are congruent.

Angle-Side-Angle (A-S-A)—If two angles and the included side of one triangle are congruent to the corresponding parts of the other triangle, the two triangles are congruent.

Angle-Angle-Side (A-A-S)—If two angles and the side opposite one of the angles in a triangle are congruent to corresponding parts of the other triangle, the triangles are congruent.

To prove that right triangles are congruent, use any method above or this method:

Hypotenuse-Leg—If the hypotenuse and either leg of one triangle is congruent to corresponding parts of the second triangle, the triangles are congruent.

LIST 103

Properties of Similar Triangles

Similar triangles have the same shape but not the same size. Three pairs of corresponding angles are congruent, and the lengths of corresponding sides are proportional. The symbol ~ means "is similar to."

Similar triangles have the following properties:

► Corresponding angles are congruent.

► Corresponding sides are in proportion.

► Two angles of one triangle are respectively congruent to two angles of the other triangle.

► An angle of one triangle is congruent to an angle of the other, and the sides including these angles are in proportion.

► An acute angle of a right triangle is congruent to an acute angle of another right triangle.

LIST 104

Proving Triangles Similar

Similar triangles have the same shape but not the same size. You can prove that two triangles are similar by using any of the methods below.

Angle-Angle (A-A)—If two angles of one triangle are congruent to corresponding angles of the other triangle, then the triangles are similar.

Side-Side-Side (S-S-S)—If the corresponding sides of one triangle are in proportion to the corresponding sides of the other triangle, then the triangles are similar.

Side-Angle-Side (S-A-S)—If a pair of corresponding angles is congruent and the sides that include these angles are in proportion, then the triangles are similar.

To prove that right triangles are similar, use any method listed above, or prove that an acute angle in one right triangle is congruent to an acute angle in the other right triangle.

Other Facts about Similar Triangles

The following have the same ratios as the lengths of any pair of corresponding sides.

▶ Corresponding altitudes

▶ Corresponding medians

▶ Corresponding perimeters

Ratio and Proportion

A *ratio* is a comparison of two values. When writing ratios, the units used for comparison must be the same. The first unit of comparison becomes the first number of the ratio. For example, since 1 yard is equal to 3 feet, the ratio of 1 foot to 1 yard can be expressed as 1 to 3.

Ratios can be expressed in three ways. A ratio of 1 to 3 may be written:

As a fraction	$\frac{1}{3}$
With a colon	1:3
With the word *to*	1 to 3

Sometimes a ratio can be simplified. The ratio $\frac{8}{10}$ can be written as $\frac{4}{5}$.

A *proportion* is an equation that states two ratios are equal. It can be written in three ways:

$$\frac{8}{10} = \frac{4}{5} \qquad 8:10 = 4:5 \qquad 8 \text{ to } 10 = 4 \text{ to } 5$$

The following facts apply to proportions.

▶ Each term of a proportion has a name determined by its position in the proportion.

▶ The first and fourth proportions are called the *extremes* of the proportion. The second and third proportions are called the *means* of the proportion.

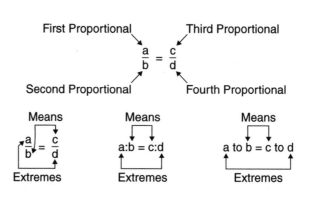

▶ In any proportion, the product of the means equals the product of the extremes. Multiplying in this way is also called *cross multiplying*. If $\frac{a}{b} = \frac{c}{d}$, then $ad = bc$.

▶ If the means of a proportion are the same, then either the second term or the third term of the proportion is called the *mean proportional* or *geometric mean* between the extremes.

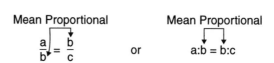

LIST 106

Deriving Proportions

From a given proportion, you may form other proportions by using some algebraic properties. The following proportions may be obtained from $\frac{a}{b} = \frac{c}{d}$, using the methods below.

▶ Switching numerators and denominators:

If $\frac{a}{b} = \frac{c}{d}$, then $\frac{b}{a} = \frac{d}{c}$ or $\frac{d}{c} = \frac{b}{a}$

▶ Switching the first proportional with the fourth proportional:

If $\frac{a}{b} = \frac{c}{d}$, then $\frac{d}{b} = \frac{c}{a}$ or $\frac{c}{a} = \frac{d}{b}$

▶ Adding the denominator to the numerator on each side of the equation:

If $\frac{a}{b} = \frac{c}{d}$, then $\frac{a+b}{b} = \frac{c+d}{d}$ or $\frac{c+d}{d} = \frac{a+b}{b}$

▶ Subtracting the denominator from the numerator on each side of the equation:

If $\frac{a}{b} = \frac{c}{d}$, then $\frac{a-b}{b} = \frac{c-d}{d}$ or $\frac{c-d}{d} = \frac{a-b}{b}$

Copyright © 2005 by Judith A. Muschla and Gary Robert Muschla

LIST 107

Principles Relating to Parallel and Proportional Lines

Parallel lines are lines that are in the same plane and never intersect. The principles listed below relate parallel lines (and in one case the angle bisector) to proportional lines.

▶ If a line is parallel to one side of a triangle, then it divides the other two sides proportionally. $\frac{a}{b} = \frac{c}{d}$

▶ If a line is parallel to one side of a triangle, then it divides the larger triangle into a smaller similar triangle. $\triangle ABC \sim \triangle ADE$

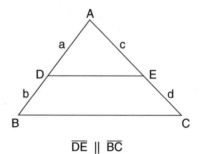

$\overline{DE} \parallel \overline{BC}$

▶ If a line divides two sides of a triangle proportionally, it is parallel to the third side.

▶ Three or more parallel lines divide any two transversals proportionally. $\frac{m}{n} = \frac{r}{s}$

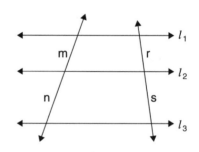

$l_1 \parallel l_2 \parallel l_3$

▶ The angle bisector of a triangle divides the opposite side into segments that are proportional to the adjacent sides.
$\frac{f}{h} = \frac{g}{j}$

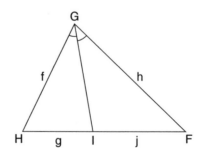

$\overline{IG}$ bisects $\angle HGF$

LIST 108

Pascal's Triangle

The triangular array of numbers below is named for Blaise Pascal, a French mathematician who lived in the seventeenth century. Each number is the sum of the numbers to its immediate upper left and immediate upper right in the previous row.

```
                1                    Row 0
              1   1                  Row 1
            1   2   1                Row 2
          1   3   3   1              Row 3
        1   4   6   4   1            Row 4
      1   5  10  10   5   1          Row 5
    1   6  15  20  15   6   1        Row 6
  .   .   .   .   .   .   .   .
```

This triangle generates many interesting patterns, some of which are noted below.

▶ Each row is symmetric; it is read the same from left to right as right to left.

▶ The triangle's line of symmetry is the vertical center line.

▶ The sum of the numbers in row n equals 2^n. For example, the sum of the numbers in row 3 is 2^3 or 8.

▶ The counting numbers are listed along the second diagonal. 1, 2, 3, 4, 5, 6 . . .

▶ The triangular numbers are listed along the third diagonal. 1, 3, 6, 10, 15 . . .

▶ The tetrahedral numbers are listed along the fourth diagonal. 1, 4, 10, 20 . . .

▶ The sum of the numbers on any diagonal is found by moving down the diagonal until you see the last number you wish to add and then moving diagonally right or left (depending on what side of the triangle you start). For example, starting from the left side of the triangle, the sum of the first five counting numbers is obtained by moving down the second diagonal until you reach the number 5, then move diagonally right to 15. Thus, the sum of the first five counting numbers is 15.

▶ The sum of all numbers above row n is $2^n - 1$. The sum of all the numbers above row 3, for instance, is $2^3 - 1$ or 7.

▶ Each row represents the coefficients in the expansion of $(x+y)^n$, where n represents the row of Pascal's triangle. For example, $(x+y)^2 = x^2 + 2xy + y^2$ has respective coefficients of 1, 2, 1, which are the numbers in row 2.

LIST **109**

The Harmonic Triangle

The *Harmonic Triangle* is a triangular array of numbers associated with Gottfried Wilhelm Leibniz (1646–1716). Considered by many to be one of the brightest men of his times, Leibniz discovered the basic principles of infinitesimal calculus.

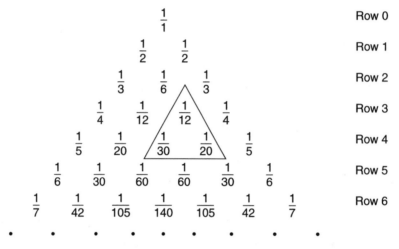

Many patterns arise from this triangle, including the following:

- ▶ Each row is symmetric. It reads the same left to right as right to left.

- ▶ The vertical center line is the line of symmetry.

- ▶ The first number in each row is the reciprocal of the counting numbers.

- ▶ The second number in each row is the product of the first number in the row and the first number in the previous row. For example, $\frac{1}{6}$ is the second number in row 2 and equals $\frac{1}{2} \times \frac{1}{3}$.

- ▶ The denominators are multiples of the numbers in Pascal's Triangle.

- ▶ An equilateral triangle can be drawn to include three numbers anywhere on the triangle with the following result: the top fraction minus the bottom left-hand fraction equals the bottom right-hand fraction—for example, in the small triangle drawn in row 3 and row 4, $\frac{1}{12} - \frac{1}{30} = \frac{1}{20}$.

- ▶ The numbers in each diagonal form an infinite sequence. For example, the terms of the sequence in the second diagonal are $\frac{1}{2}, \frac{1}{6}, \frac{1}{12}, \frac{1}{20}, \frac{1}{30}, \frac{1}{42} \ldots$.

- ▶ The sum of the infinite series whose terms are the numbers in the second diagonal equals one.

- ▶ The sum of the infinite series whose terms are the numbers in the third diagonal equals $\frac{1}{2}$.

- ▶ The sum of the infinite series whose terms are the numbers in the fourth diagonal equals $\frac{1}{3}$.

- ▶ In general, the sum of the infinite series whose terms are the numbers in the nth diagonal equals the first number (or last) in the $n - 1$ row. For example, the sum of the numbers in the fifth diagonal equals the first (or last) number in row 4, which is $\frac{1}{5}$.

LIST 110

Types of Quadrilaterals

A *polygon* is a closed plane figure whose sides are line segments, which intersect only at their endpoints. Polygons may be concave or convex, depending on the measure of their angles. Every interior angle in a convex polygon has a measure of less than 180°. At least one interior angle in a concave polygon has a measure greater than 180°.

A *quadrilateral* is a polygon that has four sides. Quadrilaterals may be subdivided into several groups that are described below.

▶ *Trapezium*—a quadrilateral with no pairs of parallel sides.

 ▪ Kite: a convex trapezium that has two congruent pairs of adjacent sides.

 ▪ Deltoid: a concave trapezium that has two congruent pairs of adjacent sides.

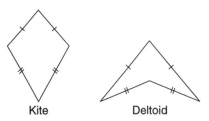

Kite   Deltoid

▶ *Trapezoid*—a quadrilateral that has only one pair of parallel sides.

 ▪ The two parallel sides are called bases.

 ▪ The legs are the sides that are not parallel.

 ▪ The median joins the midpoints of the legs.

 ▪ An isosceles trapezoid is a special type of trapezoid whose legs are congruent. They are not parallel.

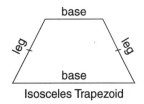

Isosceles Trapezoid

▶ *Parallelogram*—a quadrilateral whose opposite sides are parallel. This is equivalent to saying it has two pairs of parallel sides. Some special types of parallelograms follow.

 ▪ Rectangle: a parallelogram that has four right angles.

 ▪ Rhombus: a parallelogram that has four congruent sides.

 ▪ Square: a parallelogram that has four congruent sides and four right angles.

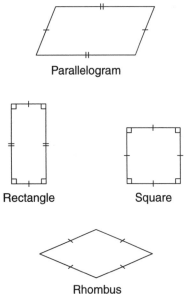

Parallelogram

Rectangle   Square

Rhombus

Properties of Special Quadrilaterals

Quadrilaterals are four-sided polygons. The sum of the interior angles of a quadrilateral is 360°. Some quadrilaterals possess special properties, as shown in the lists that follow.

Properties of a Trapezoid

▶ It possesses exactly one pair of parallel sides.

▶ The median is parallel to the bases.

▶ The length of the median equals half the sum of the length of the bases.

Properties of an Isosceles Trapezoid

▶ It possesses all the properties of a trapezoid.

▶ The lower base angles are congruent.

▶ The upper base angles are congruent.

▶ The diagonals are congruent.

▶ The legs are congruent.

Properties of a Parallelogram

▶ Opposites sides are parallel.

▶ Consecutive pairs of angles are supplementary.

▶ Opposite angles are congruent.

▶ Opposite sides are congruent.

▶ Diagonals bisect each other.

Properties of a Rectangle

▶ It possesses all of the properties of a parallelogram.

▶ Diagonals are congruent.

▶ All angles are right angles.

Properties of a Rhombus

▶ It possesses all of the properties of a parallelogram.

▶ All sides are congruent.

▶ The diagonals are perpendicular to each other.

▶ The diagonals bisect opposite angles.

▶ The slopes of the diagonals are negative reciprocals of each other.

Properties of a Square

▶ It possesses all of the properties of a parallelogram.

▶ It possesses all of the properties of a rectangle.

▶ It possesses all of the properties of a rhombus.

LIST **112**

Classification of Quadrilaterals

Quadrilaterals, which are closed figures with four sides, may be classified according to their number of parallel sides, congruent segments, and right angles. The diagram below shows the relationships of the quadrilaterals described in List 110, "Types of Quadrilaterals."

Consider this: Every square is a rectangle, but not every rectangle is a square. A rectangle, however, is also a parallelogram, but a parallelogram may not be a rectangle (unless its angles are all right angles). The next time you find yourself engaged in a conversation that is lagging, you might bring up these fine points of geometry.

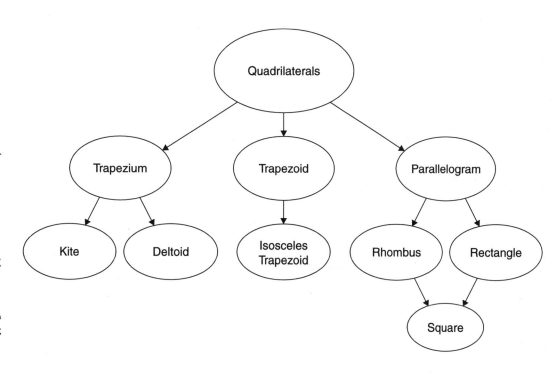

LIST 113

Proofs for Quadrilaterals

To prove that a trapezoid is isosceles, a quadrilateral is a parallelogram, and a parallelogram is a rectangle, rhombus, or square, use the following lists. Note that the lists contain the minimum information for each proof.

To prove a trapezoid is isosceles, show that at least one of the following is true:

▶ Base angles (upper or lower) are congruent.

▶ Diagonals are congruent.

▶ Legs are congruent.

To prove a quadrilateral is a parallelogram, show that at least one of the following is true:

▶ Both pairs of opposite angles are congruent.

▶ Both pairs of opposite sides are parallel.

▶ Both pairs of opposite sides are congruent.

▶ Diagonals bisect each other.

▶ One pair of opposite sides is both parallel and congruent.

To prove a parallelogram is a rectangle, show that at least one of the following is true:

▶ It contains at least one right angle.

▶ Diagonals are congruent.

To prove a parallelogram is a rhombus, show that at least one of the following is true:

▶ It contains at least one pair of congruent adjacent sides.

▶ Diagonals are perpendicular.

▶ Diagonals bisect the vertex angles.

To prove a parallelogram is a square, show that at least one of the following is true:

▶ It is a rectangle and has one pair of congruent adjacent sides.

▶ It is a rhombus and has at least one right angle.

LIST 114

Tessellations

A *tessellation* is a design that covers a plane with no gaps and no overlaps. Covering the plane in this manner is called "tessellating the plane" or "tiling the plane." Some "tessellating" facts follow.

▶ A pure tessellation uses only one shape to tile the plane.

▶ A regular tessellation uses only one regular polygon to tile the plane.

▶ A semiregular tessellation uses two or more types of regular polygons to tile the plane.

▶ Three types of polygons tile the plane to form a regular tessellation:

- square

- triangle

- hexagon

▶ Some types of polygons that form semiregular tessellations include:

- hexagon and triangle

- octagon and square

- hexagon, square, and triangle

▶ Some shapes may form a semiregular tessellation around a sphere such as hexagons and pentagons on a soccer ball.

▶ M. C. Escher (1898–1972), a Dutch artist, used tessellation in over 150 of his sketches. He used modified versions of regular tessellations by employing transformations such as translation and rotation.

Copyright © 2005 by Judith A. Muschla and Gary Robert Muschla

LIST 115

Circles

A *circle* is the set of points in a plane, all of which are the same distance from a given point. ⊙ is the symbol for a circle.

Circles are everywhere. They are represented by coins, wheels, Frisbees, and the red and black pieces used in checkers. Even a dart board has concentric circles. Following are words that apply to circles.

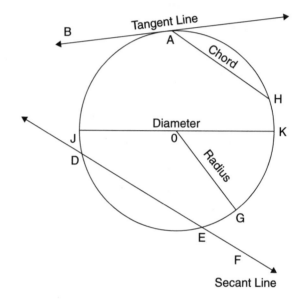

▶ *Center of a circle*—the fixed point from which all other parts are equidistant. *O* is the symbol for a circle whose center is point *O*.

▶ *Radius*—the line segment from the center of a circle to a point on the circle. $\overline{OG}$ is a radius.

▶ *Radii*—the plural form of radius.

▶ *Chord*—a line segment that connects two points on the circle. $\overline{AH}$ is a chord.

▶ *Diameter*—a chord that connects any two points on the circle through the center. It is also called the *diameteral chord*. $\overline{JK}$ is a diameter.

▶ *Circumference*—the distance around a circle. $C = \pi d$ or $C = 2\pi r$, where *C* represents the circumference, *d* represents the diameter, and *r* represents the radius.

▶ *Pi*—the ratio of the circumference to the length of the diameter. *Pi* is an irrational number approximately equal to 3.14 or $\frac{22}{7}$.

▶ *Secant*—a straight line that intersects the circle at two points. $\overleftrightarrow{DE}$ is the secant line.

▶ *Secant segment*—a line segment with one endpoint in the exterior of a circle and the other endpoint on the circle, farthest from the external point.

The secant segment is divided by the circle into two parts: the external secant segment and the internal secant segment. $\overline{DF}$ is the secant segment. $\overline{DE}$ is the internal secant segment. $\overline{EF}$ is the external secant segment.

▶ *Tangent*—a line that intersects the circle at one and only one point. This point is called the *point of tangency* or *point of contact*. $\overleftrightarrow{AB}$ is the tangent line.

▶ *Tangent segment*—a line segment that has a point on the tangent line and the point of tangency as its endpoints. $\overline{AB}$ is the tangent segment.

LIST 116

Special Circles

Based on their relative positions, two circles in a plane or a circle and a polygon have special names. Facts about these special circles follow.

- *Tangent circles*—two circles that intersect only at one point. Tangent circles may be either internally tangent or externally tangent.
  - *Internally tangent circles:* both circles are on the same side of the tangent line.
  - *Externally tangent circles:* both circles are on opposite sides of the tangent line.
- *Concentric circles*—two or more circles in a plane with the same center, but the lengths of their radii vary. The *annulus* is the region between concentric circles.
- *Eccentric circles*—circles that have different centers.
- *Circumscribed circle*—a circle passing through each vertex of a polygon.
- *Inscribed circle*—a circle to which all the sides of a polygon are tangents.
- *Circumscribed polygon*—a polygon that is outside the circle in such a way that all of its sides are tangent to the circle.
- *Inscribed polygon*—a polygon that is inside a circle so that each of its vertices lies on the circle.

Externally Tangent Circles

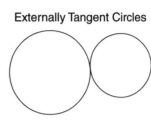

Concentric Circles

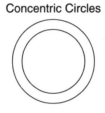

Internally Tangent Circles

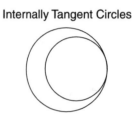

Circumscribed Circle Inscribed Polygon

Eccentric Circles

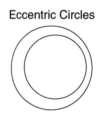

Inscribed Circle Circumscribed Polygon

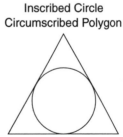

Circles: Line and Segment Principles

The lines and segments that are parts of circles have certain relationships and properties that are useful in finding lengths of line segments and congruent segments.

▶ The diameter divides a circle into two congruent parts. $\overline{AB}$ is the diameter.

▶ If a chord divides a circle into two congruent parts, the chord is a diameter.

▶ The diameter is the longest chord.

▶ The diameter equals twice the length of a radius. $\overline{OC}$ is the radius. $AB = 2(OC)$

▶ The radius equals half the length of the diameter. $OC = \frac{1}{2}(AB)$

▶ The location of points is related to the length of the radius as follows:

  ▪ A point is outside the circle if its distance from the center is greater than the length of the radius. *D* is outside the circle.

  ▪ A point is inside the circle if its distance from the center is less than the length of the radius. *E* is inside the circle.

  ▪ A point is on the circle if its distance from the center equals the length of the radius. *F* is on the circle.

▶ The radii of the same or congruent circles are equal.

▶ Diameters of the same or congruent circles are equal.

▶ A diameter perpendicular to a chord bisects the chord. $\overline{AB}$ bisects $\overline{FG}$.

▶ The radius is perpendicular to the tangent line at the point of tangency. $\overline{OC} \perp l_1$

▶ If a radius is perpendicular to a line at the point at which the line intersects a circle, then the line is a tangent.

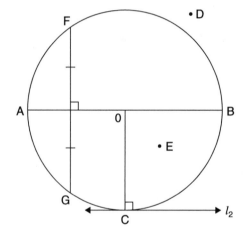

LIST 117

(Continued)

▶ A perpendicular bisector of a chord passes through the center of the circle. $\overline{AB}$ is the perpendicular bisector of $\overline{FG}$.

▶ In the same or congruent circles, congruent chords are equidistant from the center.

▶ In the same or congruent circles, chords that are equidistant from the center are congruent.

▶ If two chords intersect in the interior of a circle, then the product of the lengths of the segments of one chord equals the product of the lengths of the segments of the other chord.

▶ If two tangent segments are drawn to a circle from the same exterior point, then the tangent segments are congruent.

▶ *Secant-Secant Segment Theorem*—If two secant segments are drawn to a circle from the same exterior point, then the product of the lengths of one secant segment and its external segment is equal to the product of the lengths of the other secant segment and its external segment.

▶ *Tangent-Secant Segment Theorem*—If a tangent segment and a secant segment are drawn to the circle from the same exterior point, then the square of the length of the tangent segment is equal to the product of the lengths of the secant segment and its external segment.

$$\frac{length\ of\ arc}{circumference} = \frac{degree\ measure\ of\ arc}{360°}$$

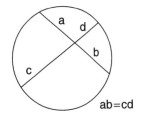

ab = cd

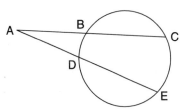

$AC \cdot BA = AE \cdot AD$

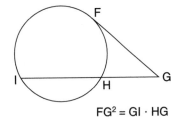

$FG^2 = GI \cdot HG$

Circles: Arcs and Angles

Just as there are special line segments that pertain to circles, there are special names for arcs and angles.

An *arc* is a part of a circle between any two points on the circle. $\overset{\frown}{AB}$ is read "arc *AB*." The slightly curved symbol above *AB* indicates an arc.

The measure of an arc is written as $m\overset{\frown}{AB}$ and is read "the measure of arc *AB*." $m\overset{\frown}{AB} = 40$ is read "the measure of arc *AB* is 40." Note that the symbol for degrees is omitted.

Types of Arcs and Angles

▶ *Semicircle*—an arc equal to half of a circle.

▶ *Minor arc*—an arc less than a semicircle. $\overset{\frown}{AC}$ is a minor arc.

▶ *Major arc*—an arc greater than a semicircle but less than 360°. $\overset{\frown}{ABC}$ is a major arc.

▶ *Midpoint of an arc*—the point of the arc that divides the arc into two congruent arcs. *B* is the midpoint of $\overset{\frown}{ABC}$ since $\overset{\frown}{AB} \cong \overset{\frown}{BC}$.

▶ *Congruent arcs*—arcs in the same or congruent circles that have the same degree measure. $\overset{\frown}{AB} \cong \overset{\frown}{AC}$.

▶ *Central angle*—an angle whose vertex is at the center of the circle; its sides are along the radii. $\angle AOC$ is a central angle.

▶ *Measure of a minor arc*—the measure of the arc's central angle. $m\overset{\frown}{AC} = n$ since $m\angle AOC = n°$.

▶ *Measure of a major arc*—360° minus the measure of the minor arc. $m\overset{\frown}{ABC} = 360° - m\overset{\frown}{AC}$.

▶ *Inscribed angle*—an angle whose vertex is on the circle and whose sides are along two chords. $\angle ABC$ is an inscribed angle.

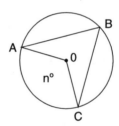

0 is the center of the circle

LIST 119

Principles of Arcs and Angles

Some important facts about arcs and angles are shown below.

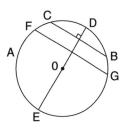

- ▶ A circle contains 360°.
- ▶ In the same or equal circles, equal arcs have equal central angles.
- ▶ In the same or equal circles, equal central angles have equal arcs.
- ▶ In the same or equal circles, equal chords have equal arcs.
- ▶ In the same or equal circles, equal arcs have equal chords.
- ▶ In the same or equal circles, parallel chords cut off equal arcs. $\overset{\frown}{CF} \cong \overset{\frown}{BG}$.
- ▶ A diameter perpendicular to a chord bisects the arcs. $\overline{ED}$ bisects $\overline{CB}$. $\overset{\frown}{DC} \cong \overset{\frown}{DB}$ and $\overset{\frown}{CFE} \cong \overset{\frown}{BGE}$.
- ▶ *Arc Addition Postulate*—If point D is on $\overset{\frown}{AB}$, then $m\overset{\frown}{AD} + m\overset{\frown}{DB} = m\overset{\frown}{AB}$.
- ▶ *Arc Sum Postulate*—If points C and D are on $\overset{\frown}{AB}$ and $m\overset{\frown}{AC} = m\overset{\frown}{BD}$, then $m\overset{\frown}{AD} = m\overset{\frown}{BC}$.
- ▶ *Arc Difference Postulate*—If points C and D are on $\overset{\frown}{AB}$ and $m\overset{\frown}{AD} = m\overset{\frown}{BC}$, then $m\overset{\frown}{AC} = m\overset{\frown}{BD}$.
- ▶ Congruent central angles intercept congruent arcs.
- ▶ Congruent arcs have congruent central angles.
- ▶ The measure of a central angle is equal to the measure of its intercepted arc.
- ▶ *Inscribed Angle Theorem*—The measure of an inscribed angle is equal to $\frac{1}{2}$ the measure of its intercepted arc. $m\angle IJK = \frac{1}{2} m\overset{\frown}{IK}$.

LIST 119

(Continued)

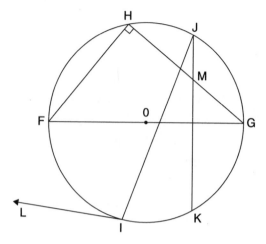

▶ An angle inscribed in a semicircle is a right angle. $\overline{FG}$ divides $\odot 0$ into two semicircles so that $\angle FHG$ is a right angle.

▶ If inscribed angles intercept the same or congruent arcs, they are congruent.

▶ If inscribed angles are congruent, their intercepted arcs are congruent.

▶ *Chord-Tangent Angle Theorem*—The measure of an angle formed by a tangent and a chord drawn from a point of tangency is equal to $\frac{1}{2}$ the measure of the intercepted arc. $\overleftrightarrow{IL}$ is tangent to $\overline{IJ}$. The measure of $\angle JIL = \frac{1}{2} m \widehat{IFG}$.

▶ *Chord-Chord Angle Theorem*—The measure of an angle formed by two chords intersecting in the interior of a circle is equal to $\frac{1}{2}$ the sum of the measures of the two intercepted arcs. $\overline{JK}$ and $\overline{HG}$ intercept at M. $m\angle HMK = \frac{1}{2}\left(m\widehat{KIH} + m\widehat{JG}\right)$.

▶ *Secant-Secant Theorem*—The measure of an angle formed by two secants intersecting in the exterior of a circle is equal to $\frac{1}{2}$ the difference of the measures of the intercepted arcs. $m\angle PQR = \frac{1}{2}\left(m\widehat{SN} - m\widehat{PR}\right)$.

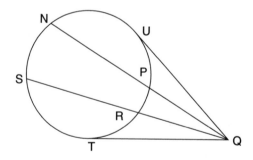

▶ *Secant-Tangent Theorem*—The measure of an angle formed by a secant and a tangent intersecting in the exterior of a circle is equal to $\frac{1}{2}$ the difference of the measures of the intercepted arcs. $m\angle RQT = \frac{1}{2}\left(m\widehat{ST} - m\widehat{RT}\right)$.

▶ *Tangent-Tangent Theorem*—The measure of an angle formed by two tangents intersecting in the exterior of a circle is equal to $\frac{1}{2}$ the difference of the measures of the intercepted arcs. $m\angle UQT = \frac{1}{2}\left(m\widehat{TSU} - m\widehat{TU}\right)$.

LIST **120**

Solids

Solid figures, also called *space figures,* are three-dimensional shapes. There are two types of solid figures:

▶ Polyhedra (singular: polyhedron) whose flat surfaces are polygons.

▶ Solid figures whose surfaces are curved.

The following vocabulary is necessary to discuss solid figures:

▶ *Face:* the flat surface of a solid.

▶ *Edge:* line segment of a solid where two faces intersect.

▶ *Vertex:* the point where the edges of a solid meet.

Types of Polyhedra

A *prism* is a polyhedron with two faces that are polygons that are both parallel and congruent. These are called the *bases.* The other faces, called *lateral faces,* are parallelograms.

Types of Prisms

▶ *Right prism*—a prism whose lateral faces are rectangles. Its bases may be any type of polygon.

Right Prism
(In this example the bases are pentagons.)

▶ *Triangular prism*—a prism that has two triangular bases and three lateral faces.

Triangular Prism

▶ *Parallelepiped*—a prism with six faces that are parallelograms.

Parallelepiped

▶ *Rectangular solid* (also called a *box*)—a parallelepiped whose faces are rectangles.

Rectangular Solid

▶ *Cube*—a rectangular solid whose faces are squares.

Cube

LIST 120

(Continued)

▶ *Pyramid*—a polyhedron that has one base. The lateral faces are triangles.

Pyramid

▶ *Regular pyramid*—a pyramid whose base is a regular polygon and whose altitude joins the vertex and the center of the base.

Regular Pyramid

Square Base

▶ *Platonic solids* (also called *regular polyhedra*)—polyhedra with congruent faces. (See List 121 for information about the five Platonic solids.)

Solid Figures That Have Curved Surfaces

The following three-dimensional figures are not polyhedra because their faces are not polygons.

▶ *Cylinder*—a solid figure with two congruent circular bases that are parallel. The line segment joining the center of the bases is an axis of the cylinder. If the axis is perpendicular to the base, then it is a right cylinder; otherwise, it is an oblique cylinder.

Cylinder

▶ *Cone*—a solid figure that has one circular base and a vertex. The line segment joining the vertex to the center of the base is an axis of the cone. If the axis is perpendicular to the base, then it is a right cone; otherwise, it is an oblique cone.

Cone

▶ *Sphere*—the set of points in space that are equidistant from the center.

Sphere

▶ A geometric solid may be produced by a plane cutting the solid parallel to the base or by two parallel planes cutting the solid. This figure is called a *frustum*.

Frustum of a Pyramid

LIST **121**

Platonic Solids

Regular polyhedra are solid figures bounded by regular polygons such that the same number of faces meet at each vertex. There are only five regular polyhedra, which are also called Platonic solids.

Name	Faces	Edges	Vertices	Types of Faces
Regular Tetrahedron	4	6	4	equilateral triangles
Regular Hexahedron or Cube	6	12	8	squares
Regular Octahedron	8	12	6	equilateral triangles
Regular Dodecahedron	12	30	20	regular pentagons
Regular Icosahedron	20	30	12	equilateral triangles

Euler's Formula: $V - E + F = 2$, where V represents the number of vertices, E represents the number of edges, and F represents the number of faces, applies to Platonic solids as well as other polyhedra.

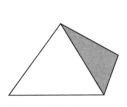

Regular Tetrahedron

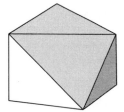

Regular Octahedron

Regular Icosahedron

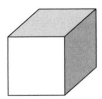

Regular Hexahedron

Regular Dodecahedron

Classifications of Solids

Solids may be grouped according to the diagram below.

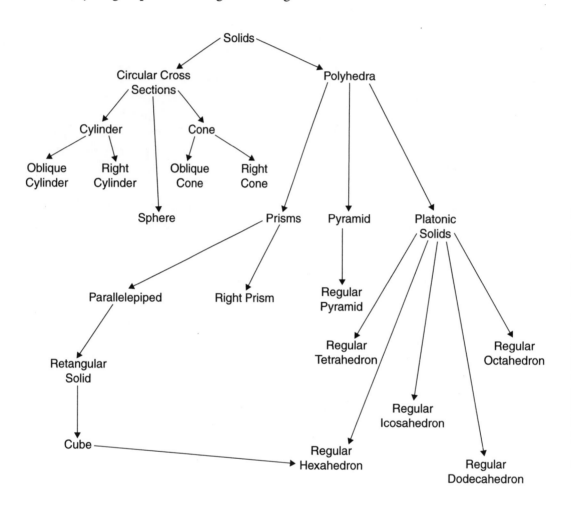

LIST 123

Patterns of a Painted Cube

By exploring the characteristics of a cube, you can see important relationships as well as some number patterns.

Suppose you dipped a cube in paint and cut it both horizontally and vertically so that the number of segments on each edge is listed in the first column of the chart below. Note the resulting number of cubes and ratios.

Number of Segments on Each Edge	Total Number of Cubes	Ratio of the Volume of Each Cube to the Volume of the Original Cube	Number of Cubes with the Given Number of Painted Sides			
			0	1	2	3
2	$2^3 = 8$	$\dfrac{1}{2^3}$ or $\dfrac{1}{8}$	0	0	0	8
3	$3^3 = 27$	$\dfrac{1}{3^3}$ or $\dfrac{1}{27}$	1	6	12	8
4	$4^3 = 64$	$\dfrac{1}{4^3}$ or $\dfrac{1}{64}$	8	24	24	8
5	$5^3 = 125$	$\dfrac{1}{5^3}$ or $\dfrac{1}{125}$	27	54	36	8
6	$6^3 = 216$	$\dfrac{1}{6^3}$ or $\dfrac{1}{216}$	64	96	48	8
7	$7^3 = 343$	$\dfrac{1}{7^3}$ or $\dfrac{1}{343}$	125	150	60	8
n	n^3	$\dfrac{1}{n^3}$	$(n-2)^3$	$6(n-2)^2$	$12(n-2)$	8

Scale Factors for Similar and Plane Figures and Solids

A *scale factor* for similar figures is the ratio of the lengths of two corresponding sides. (Similar polygons are plane figures that have the same shape but not necessarily the same size.) Corresponding angles are congruent, and corresponding sides are in proportion. The following holds true for two similar polygons.

If the scale factor is $a:b$, then

- ▶ The ratio of the perimeters is $a:b$.
- ▶ The ratio of the areas is $a^2:b^2$.

Similar solids have the same shape but not necessarily the same size. To find if two solids are similar, determine if the bases are similar and the corresponding heights are proportional. The following is true for two similar solids.

If the scale factor is $a:b$, then

- ▶ The ratio of the corresponding perimeters is $a:b$.
- ▶ The ratio of base areas is $a^2:b^2$.
- ▶ The ratio of lateral areas is $a^2:b^2$.
- ▶ The ratio of total areas is $a^2:b^2$.
- ▶ The ratio of the volumes is $a^3:b^3$.

LIST **125**

Types of Symmetry

A geometric figure is said to have symmetry (or be symmetric) with respect to a point, line, or plane if there is an exact balance in the figure. Precise definitions follow.

Symmetry with Respect to a Line

▶ Two points A and B are symmetric with respect to a line if and only if the line is the perpendicular bisector $\overline{AB}$. The line is called the *axis of symmetry* or the *line of symmetry*.

▶ A geometric figure is symmetric with respect to a line if and only if every point of the figure is balanced by a symmetrical point on the other side of the axis or line of symmetry.

Symmetry with Respect to a Plane

▶ Two points A and B are symmetric with respect to a plane if and only if the plane is both perpendicular to and bisects $\overline{AB}$. The plane is called the *plane of symmetry*.

▶ A geometric figure is symmetric with respect to a plane if and only if every point on the figure is balanced by a symmetrical point on the other side of the plane.

Symmetry with Respect to a Point

▶ Two points A and B are symmetric with respect to a point C if and only if C bisects $\overline{AB}$. The point is called the *point of symmetry*.

▶ A geometric figure is symmetric with respect to a point if and only if every point on the figure is balanced by a symmetrical point on the other side of the point.

Three of the most common types of symmetry, also known as the three basic rigid motions of geometry, are listed below:

▶ *Symmetry by reflection:* the property of being divisible into two parts that are mirror images of each other. A reflection is also called a *flip*.

▶ *Symmetry by rotation:* the property that a figure coincides with its original position when rotated about a point of symmetry through an angle of 360°. A rotation is also called a *turn*.

▶ *Symmetry by translation:* the property that a figure coincides with its original position when translated or shifted a fixed distance. A translation is also called a *glide*.

Note that some symmetries may be combinations of those listed above; for example, a glide reflection.

Types of Transformations

A *transformation,* or mapping, usually changes one quantity into another. In geometry, a transformation is the changing of one shape into another by moving each point in it to a different position, usually through a special procedure. See List 127, "Transformation Matrices." Some transformations are described below.

Isometry—a geometric transformation that preserves distance. The size and shape of the figure remain the same.

Reflection (or flip)—a geometric transformation of a point or set of points from one side of a point, line, or plane to a symmetrical position on the other side.

Rotation (or turn)—a geometric transformation in which a figure is moved about a fixed point.

Translation (or glide)—a geometric transformation in which only position relative to the axes is changed, not orientation, size, or shape.

Projection—a geometric transformation that maps a geometric figure on to a plane, producing a two-dimensional figure.

Dilation—a geometric projection in which a figure is "stretched." The resulting figure is similar to the original.

Enlargement—a dilation that produces an image larger but similar to the original shape.

Reduction—a dilation that produces an image that is smaller but similar to the original shape.

Deformation (or continuous deformation)—a geometric transformation that stretches, shrinks, or twists a shape without breaking its lines or surfaces.

Shear—a transformation that represents a shearing motion for which each point of a coordinate plane is mapped into itself.

LIST 127

Transformation Matrices

A transformation can be represented by a 2×2 matrix. Some transformation matrices follow.

Reflection in the x-axis
$$\begin{bmatrix} 1 & 0 \\ 0 & -1 \end{bmatrix}$$

Reflection in the y-axis
$$\begin{bmatrix} -1 & 0 \\ 0 & 1 \end{bmatrix}$$

Enlargement

K is the scale factor > 1
$$\begin{bmatrix} K & 0 \\ 0 & K \end{bmatrix}$$

Reduction

K is the scale factor < 1
$$\begin{bmatrix} K & 0 \\ 0 & K \end{bmatrix}$$

Stretch in the x-direction
$$\begin{bmatrix} K & 0 \\ 0 & 1 \end{bmatrix}$$

Stretch in the y-direction
$$\begin{bmatrix} 1 & 0 \\ 0 & K \end{bmatrix}$$

Shear in the x-direction by K
$$\begin{bmatrix} 1 & K \\ 0 & 1 \end{bmatrix}$$

Fractal Facts

Fractals describe complex geometric shapes that have a fractional dimension. Unlike the typical figures of geometry such as squares, rectangles, and circles, fractals can describe irregularly shaped objects. Note the following facts about fractals.

▶ Mathematician Felix Hausdorff introduced the concept of fractals in 1918.

▶ Mathematician Benoit B. Mandelbrot coined the term *fractal,* which is derived from the Latin word *fractus,* meaning "fragmented" or "broken."

▶ A fractal is an object that is both self-similar and has fractional dimension. Self-similarity means that each part is a miniature replica of the original.

▶ The fractional dimension is a way of measuring the irregularity of an object.

▶ Fractals can be formed by iteration: repeating a process over and over again, using the output of a previous procedure as the input of the next.

▶ Several fractals are named after the individuals who discovered them: George Cantor (Cantor Set), Waclaw Sierpiński (Sierpiński's Triangle), and Helge Von Koch (Koch Curve and Koch Snowflake).

▶ Fractals are used in applied mathematics in subjects as diverse as modeling systems in nature, the behavior of the stock market, and computer graphics.

LIST **129**

Formulas for Perimeter and Circumference

The *perimeter* is the distance around the edge of a polygon. The *circumference* is the distance around a circle. The formulas that follow make it easy to find perimeter and circumference.

Perimeter of a Square

▶ Multiply the length of a side by 4.

$P = 4s$

Perimeter of a Rectangle

▶ Multiply the length by 2.

▶ Multiply the width by 2.

▶ Add the products.

or

▶ Add the lengths of all the sides.

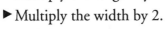

$P = 2L + 2W$

$P = 2(L + W)$

$P = L + W + L + W$

Perimeter of a Triangle

▶ Add the lengths of the sides.

$P = a + b + c$

Perimeter of a Regular Polygon (a figure such that all sides are the same length)

▶ Multiply the length of the sides by the number of sides.

$P = ns$, where
n = the number of sides and
s = the length of a side

Circumference of a Circle

▶ Multiply the length of the diameter by π. $\left(\pi \approx 3.14 \text{ or } \frac{22}{7}\right)$

or

$C = \pi d$

▶ Multiply the radius by 2.

▶ Multiply by π.

$C = 2r\pi$

LIST **130**

Area Formulas for Basic Figures

Area is the space inside a region. The following formulas can help you find the areas of various figures.

Area of a Square

▶ Multiply the length of one side by another.

$A = s^2$

Rectangle

▶ Multiply length by width.

$A = lw$

Area of a Parallelogram

▶ Multiply the length of the base by the height.

$A = bh$

Area of a Triangle

▶ Multiply $\frac{1}{2}$ by the length of the base and the height.

$A = \frac{1}{2}bh$

Area of a Trapezoid

▶ Add the lengths of parallel bases.
▶ Multiply this sum by the height.
▶ Divide the product by 2.
or

$A = \dfrac{(b_1 + b_2)h}{2}$

▶ Multiply the height by the length of the median. (The median is parallel to its bases and equals $\frac{1}{2}$ the sum of the bases.)

$A = mh$

Area of a Rhombus

▶ Multiply the lengths of the diagonals. Divide by 2.

$A = \dfrac{d_1 d_2}{2}$

Area of a Circle

▶ Square the length of the radius.
▶ Multiply by π. $\left(\pi \approx 3.14 \text{ or } \frac{22}{7}\right)$
or

$A = \pi r^2$

▶ Divide the length of the diameter by 2, square the quotient, and multiply by π.

$A = \pi \left(\dfrac{d}{2}\right)^2$

LIST 131

More Formulas for Area

Area is the space inside a given region. The following formulas can be used to find the areas of various figures.

Area of a Kite

▶ Find the product of the lengths of the diagonals.

▶ Divide by 2.

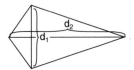

$$A = \frac{d_1 d_2}{2}$$

Area of a Triangle Using Heron's Formula

▶ Let $s = \dfrac{a+b+c}{2}$

$$A = \sqrt{s(s-a)(s-b)(s-c)}$$

Area of an Equilateral Triangle

▶ Square the length of a side.

▶ Multiply by $\sqrt{3}$.

▶ Divide by 4.

or

$$A = \frac{s^2 \sqrt{3}}{4}$$

▶ Square the height.

▶ Multiply by $\sqrt{3}$.

▶ Divide by 3.

$$A = \frac{h^2 \sqrt{3}}{3}$$

Area of a Square

▶ Square the length of a side.

or

$$A = s^2$$

▶ Square the length of a diagonal.

▶ Divide by 2.

$$A = \frac{d^2}{2}$$

Area of the Annulus (the area between concentric circles)

▶ Square the larger radius.

▶ Square the smaller radius.

▶ Subtract.

▶ Multiply by π.

$$A = \left(R^2 - r^2\right)\pi$$

LIST 131

(Continued)

Area of a Sector (the part of the circle bounded by two radii and the intercepted arc)

▶ Find the area of the circle.
▶ Find the ratio of the degrees in the central angle to 360°.
▶ Multiply the area by the ratio.

$$A = \frac{n}{360}\left(\pi r^2\right)$$

Area of a Minor Segment (the part of the circle between a chord and its arc)

▶ Find the area of the sector.
▶ Find the area of the triangle formed by the radii and the chord.
▶ Subtract.

A = area of sector − area of triangle

Area of a Regular Polygon

▶ Find the product of the number of sides of the polygon and the length of each side.
▶ Multiply the product by the apothem.
▶ Divide the product by 2.

or

▶ Find the perimeter of the polygon.
▶ Multiply the perimeter by the apothem.
▶ Divide by 2.

$A = \frac{nsa}{2}$, where n = the number of sides

$A = \frac{pa}{2}$, where p = the perimeter

Area of an Ellipse

▶ Find the product of the lengths of the minor and major axes.
▶ Divide by 4.
▶ Multiply by π.

$$A = \frac{mn}{4}\pi$$

LIST **132**

Formulas for Surface Area

The *surface area* of a space figure is the sum of the areas of all the faces of the figure. Following are the formulas for finding the surface area of several figures. Note that in all of the formulas, S stands for surface area.

Cube

$S = 6e^2$, where e = the length of an edge.

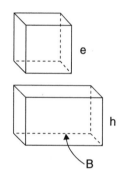

Prism

$S = 2B + Ph$, where B = the area of the base,
P = the perimeter of the base, and
h = the height of the prism.

Cylinder

$S = 2\pi r (r + h)$, where r = the radius of the circle
and h = the height of the cylinder.

Cone

$S = \pi r^2 + \pi r s$, where r = the radius of the circle
and s = the slant height.

Pyramid

$S = B + \frac{1}{2} Ps$, where B = the area of the base,
P = the perimeter of the base, and
s = the slant height of the lateral faces.

Sphere

$S = 4\pi r^2$, where r = the radius of the sphere.

LIST 133

Formulas for Finding Volume

Volume refers to the amount of space a container holds. Imagine a cardboard box, which is an example of a rectangular prism. The space inside the box is its volume. In all of the formulas below, *V* stands for volume.

Cube

$V = e^3$, where e = the length of an edge.

Rectangular Prism

$V = l \times w \times h$, where l = the length, w = the width, and h = the height of the prism.

Cylinder

$V = \pi r^2 h$ or $V = Bh$, where r = the radius of the cylinder, h = the height of the cylinder, and B = the area of the base.

Cone

$V = \dfrac{\pi r^2 h}{3}$ or $V = \dfrac{Bh}{3}$, where r = the radius of the cone, h = the height of the cone, and B = the area of the base.

Pyramid

$V = \dfrac{Bh}{3}$, where B = the area of the base, and h = the height of the pyramid.

Sphere

$V = \dfrac{4\pi r^3}{3}$, where r = the radius of the sphere.

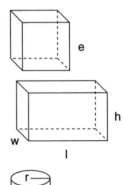

Section Four

Algebra

LIST **134**

How to Express Operations Algebraically

Changing verbal phrases into algebraic expressions is necessary to solve problems. The list below contains key words that will help you in your study of algebra.

Addition	*Subtraction*	*Multiplication*	*Division*
add	decreased by	cubed	average
augment	deduct	double	(a) fourth
combine	depreciate	factor	divided by
enlarge	difference	multiple of	equally
exceeds	diminish	multiply	half
gain	drop	quadruple	per
greater than	fewer	squared	quotient
grow	left	times	ratio
in all	less than	triple	shared
increased by	lose	twice	split
larger than	loss		(a) third
longer than	lower		
more than	minus		
plus	remain		
rise	remove		
sum	shorten		
total	smaller than		
	subtract		
	take away		

LIST **135**

Algebraic Grouping Symbols

The following symbols are important to grouping numbers, variables, and operations in algebra. Operations to be done first are enclosed in grouping symbols.

Parentheses	( )
Brackets	[ ]
Braces	{ }
Fraction Bar	——
Absolute Value	| |

LIST 136

Properties of Real Numbers

Real numbers include positive numbers, negative numbers, and zero. Since integers are a subset of the real numbers, all properties of integers are also properties of real numbers. However, some properties of real numbers are not the properties of integers.

In the chart below, *a, b,* and *c* are real numbers.

Property	Addition	Multiplication
Closure Property	$a + b$ is a unique real number	ab is a unique real number
Commutative Property	$a + b = b + a$	$ab = ba$
Associative Property	$(a + b) + c = a + (b + c)$	$(ab)c = a(bc)$
Identity Property	$a + 0 = a$	$1(a) = a$
Inverse Property	$a + (-a) = 0$	$a \times \dfrac{1}{a} = 1 \quad a \neq 0$
Property of Zero		$a(0) = 0$
Property of -1		$-1 \times a = -a$
Property of Opposites		$-(-a) = a$ $-(a + b) = -a + (-b)$ $-(ab) = (-a)b = a(-b)$
Zero Product Property		$ab = 0$ if and only if $a = 0$, $b = 0$, or both a and $b = 0$
Distributive Property	$a(b + c) = ab + ac$	
Completeness Property	Every real number can be paired with a point on the number line.	
Density Property	Between any two real numbers, there is another real number.	

Summary of Properties of Sets
of Numbers

Each set of numbers—natural numbers, whole numbers, integers, rational numbers, irrational numbers, and real numbers—has specific properties. The chart below summarizes the properties of each. A check means the property applies all the time. A blank means that a property is not always applicable.

Property	Natural	Whole	Integer	Rational	Irrational	Real
Closure (Add.)	✓	✓	✓	✓		✓
Closure (Sub.)			✓	✓		✓
Closure (Mult.)	✓	✓	✓	✓		✓
Closure (Div.)				✓		✓
Additive Identity		✓	✓	✓	✓	✓
Multiplicative Identity	✓	✓	✓	✓	✓	✓
Additive Inverse			✓	✓	✓	✓
Multiplicative Inverse				✓	✓	✓

LIST 138

Relating Operations on the Real Numbers

From addition and multiplication of real numbers (see List 136, "Properties of Real Numbers"), we can define subtraction and division. Along with each definition, other equations relating the operations of addition, subtraction, multiplication, and division follow. In the equations, *a, b, c,* and *d* are real numbers.

▶ Definition of subtraction: $a - b = a + (-b)$

▶ Definition of division: $\dfrac{a}{b} = ab^{-1} = a\left(\dfrac{1}{b}\right)$ $b \neq$

▶ $\dfrac{ac}{bc} = \dfrac{a}{b}$ $b \neq 0,\ c \neq 0$

▶ $\dfrac{a}{c} + \dfrac{b}{c} = \dfrac{a+b}{c}$ $c \neq 0$

▶ $\dfrac{a}{c} - \dfrac{b}{c} = \dfrac{a-b}{c}$ $c \neq 0$

▶ $\dfrac{a}{c} + \dfrac{b}{d} = \dfrac{ad+bc}{cd}$ $c \neq 0,\ d \neq 0$

▶ $\dfrac{a}{c} - \dfrac{b}{d} = \dfrac{ad-bc}{cd}$ $c \neq 0,\ d \neq 0$

▶ $\dfrac{a}{c} \cdot \dfrac{b}{d} = \dfrac{ab}{cd}$ $c \neq 0,\ d \neq 0$

▶ $\dfrac{a}{c} \div \dfrac{b}{d} = \dfrac{a}{c} \cdot \dfrac{d}{b} = \dfrac{ad}{cb}$ $b \neq 0,\ c \neq 0,\ d \neq 0$

▶ If $\dfrac{a}{b} = \dfrac{c}{d}$, then $ad = bc$

▶ $\dfrac{-a}{b} = \dfrac{a}{-b} = -\dfrac{a}{b}$ $b \neq 0$

▶ $\dfrac{-a}{-b} = \dfrac{a}{b}$ $b \neq 0$ $b \neq 0$

LIST 139

Axioms of Equality

An *axiom* is a self-evident principle. In algebra (and also geometry), the four following statements about equality are true for all real numbers *a*, *b*, and *c*.

Reflexive Property: $a = a$. Any number is equal to itself.

Symmetric Property: If $a = b$, then $b = a$.

Transitive Property: If $a = b$ and $b = c$, then $a = c$.

Substitution Property: If $a = b$, then *a* may replace *b* or *b* may replace *a*.

LIST 140

Axioms of Order

Just as there are general statements about equality in mathematics, there are statements about inequality. These are called *axioms of order*, also known as *axioms of inequality*.

Trichotomy Property: For all real numbers *a* and *b*, one and only one of the following statements is true: $a > b$, $a = b$, or $a < b$.

Transitive Property: For all real numbers *a*, *b*, and *c*: If $a > b$ and $b > c$, then $a > c$. If $a < b$ and $b < c$, then $a < c$.

LIST **141**

Properties of Equality

Pretend you are watching a basketball game, and the score is tied. In the next two plays, each team scores a basket. Do you agree that the score is tied once again? This is an example of the Addition Property of Equality. Try to find some examples of the other properties of equality listed below.

The following hold true when *a, b,* and *c* are real numbers.

Addition Property: If $a = b$, then $a + c = b + c$ and $c + a = c + b$. (If the same number is added to equal numbers, the sums are equal.)

Subtraction Property: If $a = b$, then $a - c = b - c$. (If the same number is subtracted from equal numbers, the differences are equal.)

Multiplication Property: If $a = b$ and $c \neq 0$, then $ac = bc$. (If equal numbers are multiplied by the same nonzero number, the products are equal.)

Division Property: If $a = b$ and $c \neq 0$, then $\frac{a}{c} = \frac{b}{c}$. (If equal numbers are divided by the same nonzero number, the quotients are equal.)

Properties of Inequalities

Inequalities are mathematical sentences that show a relationship between two or more variables. The following signs are used to show inequalities:

less than: <

less than or equal to: ≤

greater than: >

greater than or equal to: ≥

is not equal to: ≠

The following properties of inequalities hold true for all real numbers a, b, c, and d.

Addition

If $a > b$, then $a + c > b + c$.
If $a < b$, then $a + c < b + c$.
If $a < b$ and $c < d$, then $a + c < b + d$.
If $a > b$ and $c > d$, then $a + c > b + d$.

Subtraction

If $a > b$, then $a - c > b - c$.
If $a < b$, then $a - c < b - c$.

Multiplication*

If $a > b$ and $c > 0$, then $ac > bc$.
If $a > b$ and $c < 0$, then $ac < bc$.
If $0 < a < b$ and $0 < c < d$, then $ac < bd$.

Division*

If $a > b$ and $c > 0$, then $\frac{a}{c} > \frac{b}{c}$.

If $a > b$ and $c < 0$, then $\frac{a}{c} < \frac{b}{c}$.

If $a < b$ and $ab > 0$, then $\frac{1}{a} > \frac{1}{b}$.

*For multiplication and division, the above properties are not valid if $c = 0$.

LIST 143

Powers of Real Numbers

Some numbers may be written as the product of numbers that have identical factors—for example, $100 = 10 \times 10$ or 10^2. In this case, 10 is called the *base*, and 2 is the *exponent*. The exponent shows the number of times the base is a factor.

Powers of a number can be written in factored form or in exponential form:

▶ Factored form indicates the products of the factors as in $b \cdot b \cdot b \cdot b \cdot b$.

▶ Exponential form indicates the base and exponents as in b^5.

Examples of some of the powers of the real number b follow.

	Factored Form	Exponential Form	Read
Zero power of b	$1(b \neq 0)$	b^0	1
First power of b	b	b^1 or b	b to the first power
Second power of b	$b \cdot b$	b^2	b to the second power, b squared, or the square of b
Third power of b	$b \cdot b \cdot b$	b^3	b to the third power, b cubed, or the cube of b
Fourth power of b	$b \cdot b \cdot b \cdot b$	b^4	b to the fourth power, or b to the fourth
nth power of b (n is a positive integer that represents the number of times b is multiplied)	$b \cdot b \cdot b \cdot b \ldots b$	b^n	b to the nth power, or b to the nth
$-n$th power of b (n is a positive integer that represents the number of times b is multiplied. $b \neq 0$)	$\dfrac{1}{b \cdot b \cdot b \cdot b \ldots b}$	b^{-n}	b to the $-n$th power, or b to the $-n$th

LIST 144

Rules for Exponents

The following rules for exponents hold for real numbers a and b. m and n are rational numbers.

For Multiplication: $a^m \cdot a^n = a^{m+n}$

For Division: $\dfrac{a^m}{a^n} = a^{m-n}$ $a \neq 0$

For a Power of a Power: $\left(a^m\right)^n = a^{mn} = \left(a^n\right)^m$

For a Power of a Product: $(ab)^m = a^m b^m$

For a Power of a Quotient: $\left(\dfrac{a}{b}\right)^m = \dfrac{a^m}{b^m}$ $b \neq 0$

For a Zero Exponent: $a^0 = 1$ $a \neq 0$

For an Exponent of 1: $a^1 = a$

For a Negative Exponent: $a^{-n} = \dfrac{1}{a^n}$ $a \neq 0$

For a Base of 1: $1^n = 1$

LIST 145

Order of Operations

In algebra, finding the right answer often depends on the way you go about solving a problem. There is a specific order of operations you must follow.

1. Simplify expressions within grouping symbols. If several grouping symbols are used, simplify the innermost group first, and continue simplifying to the outermost group. As you do, be sure to follow steps 2, 3, and 4.

2. Simplify powers.

3. Perform all multiplicative operations (multiplication and division) from left to right.

4. Perform all additive operations (addition and subtraction) from left to right.

LIST 146

How to Construct a Number Line

Use the following steps to construct number lines accurately.

1. Use a ruler to draw a line segment.
2. Show continuity by drawing arrows on both ends.
3. Using a ruler, divide the line into equal segments.
4. Pair the endpoints or successive endpoints with integers listed in chronological order. Be sure to label numbers to the left of zero with a negative sign.

Steps

1)

2)

3)

4)

−5 −4 −3 −2 −1 0 1 2 3 4 5

LIST 147

Steps for Graphing on a Number Line

The following list provides simple procedures for graphing points and inequalities on number lines.

To Graph a Point

1. Locate the coordinate (number) on the number line.
2. Place a shaded circle on the number line above the coordinate.

   Example: $x = 5$.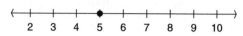

To Graph an Inequality

1. Locate the coordinate (number) on the number line.
2. If $x >$ a number, circle that number on the number line and shade the number line to the right.
3. If $x \geq$ a number, place a shaded circle on the number line and shade the number line to the right.
4. If $x <$ a number, circle that number on the number line and shade the number line to the left.
5. If $x \leq$ a number, place a shaded circle on the number line and shade the number line to the left.
6. If $x \neq$ a number, circle that number on the number line and shade to left and right.

   Examples:

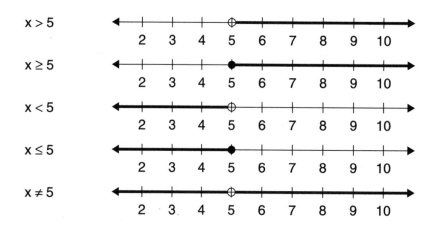

LIST 148

The Absolute Facts on Absolute Value

Absolute value is the distance a number is from zero on the number line.

$$|a| = a \text{ if } a \geq 0$$
$$|a| = -a \text{ if } a < 0$$

The distance between two real numbers a and b on the number line can be found by $|a - b|$ or $|b - a|$.

To evaluate an expression using absolute value, evaluate the expression within the absolute value symbols first. The symbols have the same priority as parentheses in the order of operations.

The absolute value of a sum is less than or equal to the sum of the absolute values.

$$|a + b| \leq |a| + |b|$$

The absolute value of a product is the product of the absolute values.

$$|ab| = |a||b|$$

If the absolute value of an expression is greater than a positive number, a *disjunction* results.

For a real number $c > 0$,
$|a| > c$ is equivalent to $a < -c$ or $a > c$

If the absolute value of an expression is less than a positive number, a *conjunction* results.

For a real number $c > 0$,
$|a| < c$ is equivalent to $-c < a < c$ or $-c < a$ and $a < c$

If the absolute value of an expression is less than a negative number, there is no solution. The absolute value is always greater than or equal to zero and therefore is greater than any negative number—for example: $|n| < -4 \quad n = \varnothing$ (no solution).

If the absolute value of an expression is greater than a negative number, the solution set is all real numbers. The absolute value is always greater than or equal to zero and therefore is always greater than a negative number—for example: $|n| > -4 \quad n =$ all numbers.

Steps to Solve an Equation in One Variable

Equations that have the same solution are called *equivalent equations*. The following is a step-by-step list for rewriting and transforming the original equation into an equivalent equation that has the same solution or root. If a step does not apply to a specific problem, go on to the next step.

1. Simplify each side of the equation. This may include:
   - ▶ Combining similar terms within grouping symbols
   - ▶ Using the Distributive Property
   - ▶ Removing any unnecessary parentheses
   - ▶ Combining similar terms

2. Add (or subtract) the same real number to (or from) each side of the equation. (If you add or subtract zero, an equivalent equation will result. It will be the same as the previous equations and will not be easier to solve. Although you can add or subtract zero, it is an unnecessary step and should be avoided.)

3. Multiply (or divide) each side of the equation by the same nonzero real number. (If you multiply each side of the equation by zero, the result will always be $0 = 0$, and the equation would not be solved. You cannot divide each side of an equation by zero because division by zero is undefined.)

4. In most cases, there is only one solution. The final transformation will result with the variable equaling a real number.

5. If the final transformation is equivalent to a false statement such as $3 = 7$ or $0 = 8$, the equation has no solution or root. It is written as $\emptyset$.

6. If the final transformation is equivalent to a statement that is always true such as $x = x$ *or* $3 = 3$, the equation is called an *identity* and is true for all real numbers.

LIST 150
Steps to Solve an Inequality in One Variable

Just as equivalent equations have the same solution set, equivalent inequalities have the same solutions. To solve an inequality, try to rewrite it, and transform it into an equivalent inequality using many of the same steps you use to solve equations. Be careful, however, when you multiply or divide both sides of the inequality by the same negative number, because that reverses the direction of the inequality. To solve an inequality in one variable, follow the steps below. If a step does not apply to a specific problem, go on to the next step.

1. Simplify each side of the inequality. This may include:
   ▶ Combining similar terms within grouping symbols
   ▶ Using the Distributive Property
   ▶ Removing any unnecessary parentheses
   ▶ Combining similar terms

2. Add (or subtract) the same real number to (or from) each side of the inequality. Adding or subtracting zero should be avoided.

3. Multiply (or divide) each side of the inequality by the same positive real number.

4. Multiply (or divide) each side of the inequality by the same negative real number, and reverse the direction of the inequality.

5. In most cases, the final transformation will be a comparison of a variable and real number, such as $x > 7$.

6. If the final transformation is equivalent to a false statement, such as $2 > 7$, $6 > 9$, or $x < x$, the inequality has no solution or root.

7. If the final transformation is equivalent to a statement that is always true, such as $x \geq x$ or $4 < 5$, the inequality is true for all real numbers.

Polynomials

A *monomial* is an expression that is either (1) a real number, (2) a variable, or (3) the product of a real number and one or more variables. Remember that a variable cannot be in the denominator. The degree of a monomial is the sum of the exponents of its variables.

A *polynomial* is the sum or difference of monomials. Monomials that make up a polynomial are called its *terms*. Polynomials of two or three terms have special names, as shown below.

> ▶ *Binomial:* a polynomial with two terms.
> ▶ *Trinomial:* a polynomial with three terms.

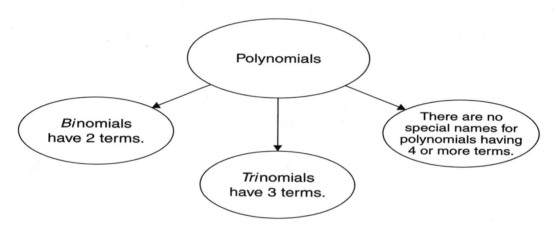

To simplify a polynomial, combine similar terms. Similar terms contain the same variables and same exponents. Simplified polynomials can be arranged in the following orders:

> ▶ Descending order in which the degree of each term decreases in successive terms.
> ▶ Ascending order in which the degree of each term increases in successive terms.

The degree of a polynomial is the highest degree of any of its terms after it has been simplified.

LIST 152

Multiplication with Monomials and Polynomials

To multiply a polynomial by a monomial, use the distributive property.

$$a(b + c + d) = ab + ac + ad$$

To multiply two binomials, use the FOIL method.

$$(a + b)(c + d) = ac + ad + bc + bd$$

FOIL is an acronym for:

F—product of the FIRST terms	ac
O—product of the OUTERMOST terms	ad
I—product of the INNERMOST terms	bc
L—product of the LAST terms	bd

If the outermost and innermost products can be simplified, do so.

Special cases:

$$c(a + b) = ca + cb$$
$$c(a - b) = ca - cb$$
$$(a + b)^2 = a^2 + 2ab + b^2$$
$$(a - b)^2 = a^2 - 2ab + b^2$$
$$(a - b)(a + b) = a^2 - b^2$$
$$(a + c)(a + d) = a^2 + (c + d)a + cd$$

Guidelines for Factoring Polynomials of Degree 2

A polynomial is factored completely when it is written as the product of a prime polynomial and monomial, or it is the product of prime polynomials. A *prime polynomial* is a polynomial that cannot be factored. Use the following suggestions for factoring polynomials completely.

► Factor out the greatest monomial factor (GMF). The GMF is the largest monomial that is a factor of each term in the polynomial.

► If the polynomial has two terms, look for the difference of squares.

► If the polynomial has three terms, look for a perfect square trinomial or a pair of binomial factors.

► If the polynomial has four or more terms, group the terms, if possible, in ways that can be factored. Factor out common polynomials.

► Be sure each polynomial is prime.

► Check by multiplying all factors.

► Remember that not all polynomials can be factored.

For additional information, see List 154, "Common Factoring Formulas."

LIST 154

Common Factoring Formulas

While there are many ways to factor polynomials, the use of formulas can be very helpful. The list below contains formulas for factoring polynomials of degree 2 or higher.

Factoring the Greatest Common Factor

$$ca + cb = c(a + b)$$
$$ca - cb = c(a - b)$$

Difference of Squares

$$a^2 - b^2 = (a - b)(a + b)$$

Sum of Squares

$$a^2 + b^2 \quad \text{prime polynomial (cannot be factored over real numbers)}$$

Perfect Square Trinomials

$$a^2 + 2ab + b^2 = (a + b)(a + b) = (a + b)^2$$
$$a^2 - 2ab + b^2 = (a - b)(a - b) = (a - b)^2$$

Other Polynomials

$$a^2 + (c + d)a + cd = (a + c)(a + d)$$
$$a^3 + 3a^2b + 3ab^2 + b^3 = (a + b)^3$$
$$a^3 - 3a^2b + 3ab^2 - b^3 = (a - b)^3$$
$$a^4 - 4a^3b + 6a^2b^2 + 4ab^3 + b^4 = (a - b)^4$$
$$a^3 - b^3 = (a - b)(a^2 + ab + b^2)$$
$$a^3 + b^3 = (a + b)(a^2 - ab + b^2)$$
$$1 - a^n = (1 - a)(1 + a + a^2 + \ldots + a^{n-1})$$
$$a^n - b^n = (a - b)\left(a^{n-1} + ba^{n-2} + b^2a^{n-3} + \ldots + b^{n-2}a + b^{n-1}\right) \text{ for } n \text{ positive and even}$$
$$a^n + b^n = (a + b)\left(a^{n-1} - ba^{n-2} + b^2a^{n-3} - \ldots - b^{n-2}a + b^{n-1}\right) \text{ for } n \text{ positive and odd}$$

Characteristics of the Coordinate Plane

The *coordinate plane* may be thought of as a flat surface divided into four parts, or quadrants, by the intersection of a vertical number line (called the *y*-axis) and a horizontal number line (called the *x*-axis). The coordinate plane is used to graph ordered pairs of the form (*x,y*), straight lines, and other functions or relations. The *x*-coordinate of the ordered pair is called the *abscissa*. The *y*-coordinate of the ordered pair is called the *ordinate*. A coordinate plane and some of its most important characteristics are shown below.

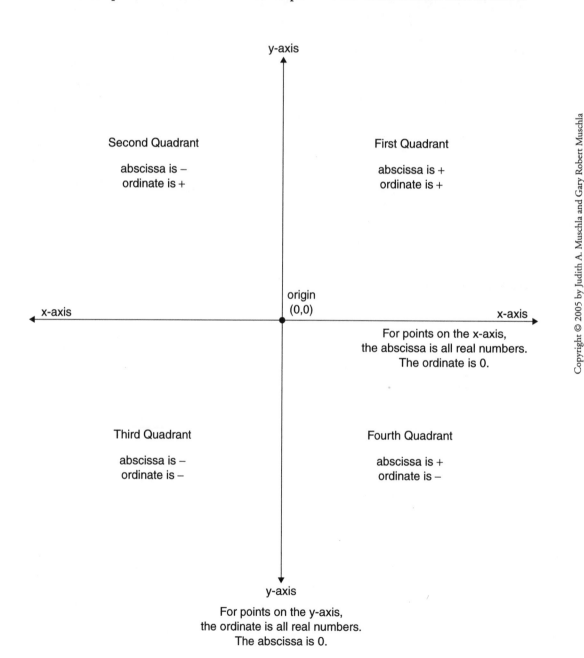

LIST 156
Plotting Points on the Coordinate Plane

Plotting, or graphing, points on the coordinate plane involves moving horizontally and vertically, depending on the values of x and y. The following "points" will help you to graph correctly.

1. In the ordered pair (x,y), the first coordinate (called the abscissa) is the value of x. Start from the origin. If the abscissa is:
   - 0, remain at the origin.
   - positive, move to the right the required number of spaces along the x-axis and stop.
   - negative, move to the left the required number of spaces along the x-axis and stop.

2. The second coordinate (called the ordinate) is the value of y. If the ordinate is:
   - 0, do not move up or down from the point where your pencil stopped after finding the abscissa. Graph the point by making a dot.
   - positive, move directly up the required number of spaces from where your pencil stopped after finding the abscissa. Graph the point by marking a dot.
   - negative, move directly down the required number of spaces from where your pencil stopped after finding the abscissa. Graph this point by marking a dot.

3. Label the point by writing the coordinates near the point.

   Following are some special points.
   - $(0,0)$ is the origin.
   - $(0,y)$ is a point on the y-axis, provided y is a real number.
   - $(x,0)$ is a point on the x-axis, provided x is a real number.
   - When $x > 0$ and $y > 0$, (x,y) is a point in the first quadrant.
   - When $x < 0$ and $y > 0$, (x,y) is a point in the second quadrant.
   - When $x < 0$ and $y < 0$, (x,y) is a point in the third quadrant.
   - When $x > 0$ and $y < 0$, (x,y) is a point in the fourth quadrant.

Common Forms of Linear Equations

A *linear equation* is an equation whose graph is a straight line. It is an equation of the first degree and is usually expressed in one of the following forms.

$ax + by = c$ **Standard Form**
a, *b*, and c are integers.
Both *a* and *b* cannot equal zero.

$x = k$ **Vertical Line**
k is any real number.
Vertical lines have no slope.

$y = k$ **Horizontal Line**
k is any real number.
Horizontal lines have a slope of zero.

$y = mx + b$ **Slope-Intercept Form**
m stands for the slope.
b stands for the *y*-intercept.

$y = mx$ **Slope-Intercept Form of a Line**
        Passing Through the Origin
m stands for the slope.

$y - y_1 = m(x - x_1)$ **Point-Slope Form**
m stands for the slope.
(x_1, y_1) is a point on the line.

LIST 158

Formulas and the Coordinate Plane

Some equations are frequently used to graph lines or points on the coordinate plane. They are summarized below.

► Slope of a line given two points (x_1, y_1) and (x_2, y_2).

- $m = \dfrac{y_2 - y_1}{x_2 - x_1}$
- m stands for the slope.
- Horizontal lines have a slope of 0.
- Vertical lines have no slope.

► Slope-Intercept Equation: $y = mx + b$.

- m stands for the slope.
- b stands for the y-intercept.

► Standard form of a linear equation: $ax + by + c = 0$.

- a, b, and c are integers.
- Both a and b cannot equal 0.

► Point-Slope Form: $y - y_1 = m(x - x_1)$.

- m stands for the slope.
- (x_1, y_1) is a point on the line.

► Distance Formula: $d = \sqrt{(x_2 - x_1)^2 + (y_2 - y_1)^2}$

- d is the distance between two points, (x_1, y_1) and (x_2, y_2).

► Midpoint Formula: $\left(\dfrac{x_1 + x_2}{2}, \dfrac{y_1 + y_2}{2} \right)$

- This formula gives the coordinate of the point halfway between (x_1, y_1) and (x_2, y_2).

► Parallel lines have the same slope.

- $m_1 = m_2$ if and only if l_1 and l_2 are two nonvertical, noncollinear straight lines with slopes m_1 and m_2.

► The slopes of perpendicular lines are negative reciprocals of each other.

- $m_1 \cdot m_2 = -1$, m_1, $m_2 \neq 0$ if and only if l_1 and l_2 are nonvertical, nonhorizontal straight lines with slopes m_1 and m_2.

LIST 159

Graphing Linear Equations in Two Variables on the Coordinate Plane

You can graph linear equations in two variables on the coordinate plane in three ways: plotting points, using intercepts, or using the slope-intercept method.

Plotting Points

1. Find two ordered pairs that satisfy the equation. This can be done by looking at the equation, or by choosing an x value (choose a value that will simplify the arithmetic), and then finding a corresponding y value.

2. Plot these points on the coordinate plane.

3. Draw a straight line through the points. This line is the graph of the equation.

4. Check the accuracy of your graph by plotting a third point. This point should be on the line. If it is not, go back and check your work and graph. All points should be collinear (lie on the same line).

Using Intercepts

1. Transform the equation into the form $Ax + By = C$.

2. Substitute 0 for y to find the x-intercept.

3. Substitute 0 for x to find the y-intercept.

4. Plot the intercept points.

5. Draw a straight line through the points. This is the graph of the line.

Using the Slope-Intercept Method

1. Transform the equation into the form of $y = mx + b$. If there is no y term, the graph is a vertical line. Solve for x. This is the x-intercept.

2. Using the equation $y = mx + b$, and, assuming there is a y term, graph the point $(0,b)$. This is the y-intercept.

3. Write the slope m as a fraction. Remember that $m = \frac{rise}{run}$.

4. From the y-intercept, count out the rise and run. Graph this point. If the rise is positive, count up. If the rise is negative, count down. If the run is positive, count to the right. If the run is negative, count to the left.

5. Draw a straight line through the y-intercept and the point plotted by using the slope. This is the graph of the line.

LIST 160

Graphing a Linear Inequality in Two Variables on the Coordinate Plane

The following steps and table will help you to graph linear inequalities in two variables on coordinate planes.

1. Graph the inequality as if it were an equation. This will enable you to find the boundary, which is the line that divides the coordinate plane into two half-planes. (You may find it helpful to refer to List 159, "Graphing Linear Equations in Two Variables on the Coordinate Plane.")

2. Graph the line. If the inequality symbol is ≥ or ≤, draw a solid line since these solutions are included. If the inequality symbol is > or <, draw a broken line since these solutions are not included.

3. Choose a point in the plane that is not on the line. Substitute the coordinates of this point in the inequality.

4. If the inequality is true, shade the half-plane that includes the point. If the inequality is false, shade the other half-plane.

The following table offers some guidelines.*

Equation	Type of Line	Shaded
$x \geq k$	Solid Vertical	Right
$x > k$	Broken Vertical	Right
$x \leq k$	Solid Vertical	Left
$x < k$	Broken Vertical	Left
$y \geq k$	Solid Horizontal	Above
$y > k$	Broken Horizontal	Above
$y \leq k$	Solid Horizontal	Below
$y < k$	Broken Horizontal	Below
$y \geq mx + b$	Solid	Above
$y > mx + b$	Broken	Above
$y \leq mx + b$	Solid	Below
$y < mx + b$	Broken	Below

*k stands for any real number, m stands for the slope, and b stands for the y-intercept.

Steps to Solve a System of Linear Equations in Two Variables

There are four methods to solve a system (more than one) of linear equations. Although any method can be used, some may be more efficient and direct for certain systems than others. A list of methods and how they can best be used follows.

Graphing Method

1. Draw the graph of each equation on the same coordinate plane. The lines will either intersect at one point, be parallel, or coincide.
2. If the lines intersect, the coordinates of the point of intersection are the solution to the two equations.
3. If the lines are parallel, there is no solution.
4. If the lines coincide (that is, the lines are identical), each point on the line is a solution. The number of solutions is infinite.

Use this method when you wish to approximate the solution. It is also most helpful when the solution is near the origin. This method is used the least for solving systems of linear equations.

Substitution Method

1. Solve one equation for one of the variables whose coefficient is one.
2. Substitute this expression in the equation you have not used. You should now have an equation in one variable. Solve this equation.
3. Substitute this expression in the equation you used in Step 1, and solve it.
4. Check your answers in both original equations.

Use this method when the coefficient of one of the variables is 1 or −1.

Addition-or-Subtraction Method

1. Add or subtract equations to eliminate one variable. Add the equations if the coefficients of one of the variables are opposites; subtract if the coefficients of one of the variables are the same.
2. Solve the equation resulting from Step 1.
3. Substitute this value in either of the original equations.
4. Check your answers in both of the original equations.

Use this method if the coefficients of one of the variables are the same or if the coefficients are opposites.

LIST 161
(Continued)

Multiplication with Addition-or-Subtraction Method

1. Multiply one or both equations so that the coefficients of one of the variables will be the same or opposite.
2. Follow steps 1 through 4 in the Addition-or-Subtraction Method.

Use this method for the following conditions:

▶ To clear the equations of fractions

▶ If the coefficients of a variable are relatively prime (have the greatest common factor of 1)

▶ If one of the coefficients of a variable is a factor (other than 1) of the other

LIST 162

Types of Functions

A *function* is a special type of relation in which every element in the domain is paired with exactly one element of the range. This is loosely translated as: "For each value of *x*, there is one and only one *y*." As a counterexample, if there are two or more *y* values for any value of *x*, the relation is not a function.

If a vertical line can be drawn anywhere on a graph and the vertical line intersects the graph at more than one point, then the graph is not the graph of a function. This is called the Vertical Line Test.

Following is a list of functions and their descriptions.

▶ Linear: $y = mx + b$ or $f(x) = mx + b$ where *m* and *b* are real numbers (*m* stands for the slope and *b* stands for the *y*-intercept). Special names for linear functions include:

- Constant Function: $y = b$ or $f(x) = b$.

  The slope is zero.

  The graph is a horizontal line.

- Identity Function: $y = x$ or $f(x) = x$.

  The slope is 1.

  The graph is a line passing through the origin.

- Direct Variation: $y = mx$ or $f(x) = mx$.

  The slope is not equal to zero.

  The graph is a line passing through the origin.

▶ Absolute Value Function: $y = |x|$ or $f(x) = |x|$.

- If $x \geq 0$, the graph is like the graph of $y = x$.
- If $x < 0$, the graph is like the graph of $y = -x$.

▶ Greatest Integer Function: $y = \left[\,|x|\,\right]$ or $f(x) = \left[\,|x|\,\right]$.

- This graph finds the greatest integer that is not greater than *x*.
- The graph is a series of line segments with one open endpoint.

▶ Inverse Variation Function: $xy = k$ or $f(x) = \dfrac{k}{x}$; $k \neq 0, x \neq 0$.

- The graph is a hyperbola.

▶ Quadratic Function: $y = ax^2 + bx + c$ or $f(x) = ax^2 + bx + c$; $a \neq 0$.

- The graph is a parabola.

▶ Cubic Function: $y = ax^3 + bx^2 + cx + d$ or $f(x) = ax^3 + bx^2 + cx + d$; $a \neq 0$.

- The graph resembles a sideways *S*.

LIST 162

(Continued)

▶ Exponential Function: $y = b^x$ or $f(x) = b^x$; $\ b > 0$, $b \neq 1$, $\ x$ is a real number.
  - The graph resembles part of a hyperbola.

▶ Logarithmic Function: $y = \log_a x$ if and only if $a^y = x$; $\ a > 0$, $a \neq 0$.
  - This function is the inverse of the Exponential Function.
  - The graph is the inverse of the graph of the Exponential Function.

Note: For Trigonometric Functions, see Lists 183 and 186 for definitions.

Some functions may be classified as odd or even. The properties of each are shown below:

▶ Even Function: $y = f(-x) = f(x)$ for all x in the domain.
  - The graph is symmetric with respect to the y-axis. [If (x,y) is on the graph, then so is $(-x,y)$.]
  - *Examples:* $y = |x|$, $y = x^2$

▶ Odd Function: $y = f(-x) = -f(x)$ for all x in the domain.
  - The graph is symmetric with respect to the origin. [If (x,y) is on the graph, then so is $(-x, -y)$.]
  - *Examples:* $y = x$, $y = x^3$

Direct Facts on Variation

Some functions are used so frequently in science and math that they have special names and general formulas. Types of "variations" fall into this category.

Common Types of Variations*

Formula	Meaning
$y = kx$ $\dfrac{y_1}{x_1} = \dfrac{y_2}{x_2}$	y varies directly as x or y is directly proportional to x.
$y = kx^2$ $\dfrac{y_1}{x_1^2} = \dfrac{y_2}{x_2^2}$	y varies directly as the square of x or y is directly proportional to the square of x.
$y = kx^3$ $\dfrac{y_1}{x_1^3} = \dfrac{y_2}{x_2^3}$	y varies directly as the cube of x or y is directly proportional to the cube of x.
$y = \dfrac{k}{x}$ $xy = k$ $x_1 y_1 = x_2 y_2$	y varies inversely as x or y is inversely proportional to x.
$y = \dfrac{k}{x^2}$ $x^2 y = k$ $x_1^2 y_1 = x_2^2 y_2$	y varies inversely as the square of x or y is inversely proportional to the square of x.
$z = kxy$ $\dfrac{z_1}{x_1 y_1} = \dfrac{z_2}{x_2 y_2}$	z varies jointly as x and y.
$z = \dfrac{kx}{y}$ $zy = kx$ $\dfrac{z_1 y_1}{x_1} = \dfrac{z_2 y_2}{x_2}$	z varies directly as x and inversely as y. This is a combined variation.

*k is a nonzero constant. It is called the *constant of variation* or the *constant of proportionality.*

LIST 164

Functional Facts About Functions

Functions are sometimes combined to form other sums, differences, products, quotients, and inverses. Following are some functional facts.

If f and g are any two functions with a common domain, then:

1. The sum of f and g, written $f + g$, is defined by $(f + g)(x) = f(x) + g(x)$.

2. The difference of f and g, written $f - g$, is defined by $(f - g)(x) = f(x) - g(x)$.

3. The product of f and g, written fg, is defined by $(fg)(x) = f(x)g(x)$.

4. The quotient of f and g, written $\dfrac{f}{g}$, is defined by $\left(\dfrac{f}{g}\right)(x) = \dfrac{f(x)}{g(x)}$. $g(x) \neq 0$.

5. The composite of f and g, written $f \circ g$, is defined by $[f \circ g](x) = f[g(x)]$.

6. To find the inverse of a function, interchange the values for x and y and then solve for y. By doing this, the order of each pair in f is reversed. (Note that the inverse of a function is not always a function.)

7. Assume f and f^{-1} are inverse functions. Then $f(a) = b$ if and only if $f^{-1}(b) = a$.

8. f and f^{-1} are inverse functions if and only if their composites are the identity function,* meaning $[f \circ f^{-1}](x) = [f^{-1} \circ f](x) = x$.

---

*The identify function is $y = x$ or $f(x) = x$. See List 162, "Types of Functions."

Square Roots

Finding the square root of a number is the inverse of squaring a number. Since $5^2 = 25$ and $(-5)^2 = 25$, the square root of 25 is both 5 and -5. Square root notation and properties follow.

- $\sqrt{}$ is called the radical sign.

- The number written beneath the radical sign is called the radicand. *Example*: $\sqrt{a}$.

- $\sqrt{a}$ is used to denote the principal or nonnegative square root of a positive real number a.

- $-\sqrt{a}$ is used to denote the negative square root of a positive real number a.

- $\pm\sqrt{a}$ is used to denote the positive or negative square root of a positive real number a.

- The index of a root is the small number written above and to the left of the radical sign. It represents which root is to be taken. The index for square roots is 2. It is understood and therefore not included.

- $\left(\sqrt{a}\right)^2 = a$ where a is a positive real number.

- $\sqrt{a^2} = |a|$ where a is a real number.

- $\sqrt{a^{2m}} = |a|^m$ where a is any real number, and m is any positive integer.

- 0 has only one square root, $\sqrt{0} = 0$.

- Negative numbers do not have square roots in the set of real numbers. See List 170, "Imaginary Numbers and Their Powers."

- In accordance with the Product Property of Square Roots, $\sqrt{ab} = \sqrt{a}\sqrt{b}$, where a and b are nonnegative real numbers.

- In accordance with the Quotient Property of Square Roots, $\sqrt{\dfrac{a}{b}} = \dfrac{\sqrt{a}}{\sqrt{b}}$, where a is any nonnegative real number and b is a positive real number.

- In accordance with the Property of Square Roots of Equal Numbers, $a^2 = b^2$ if and only if $a = b$ or $a = -b$, where a and b are real numbers.

- A radical is in simplest form when:
  - No radicand has a square root factor (other than 1).
  - The radicand is not a fraction.
  - No radicals are in the denominator.

- Only radicals with like radicands may be added or subtracted.

LIST 166

Nth Roots

If the volume of a cube is 125 cubic units, then any side of the cube has a length of 5 units. You can write $\sqrt[3]{125} = 5$ because $5^3 = 125$. We can also say $\sqrt[3]{a} = b$, if $b^3 = a$.

The small number written in the upper left-hand corner of the radical is called the *index*. When finding the square root of a number, the index is generally not stated. It is understood to be 2.

For any integer $n \geq 2$, the *n*th root is defined as follows:

$\sqrt[n]{a} = b$, if and only if $b^n = a$

for $a \geq 0$ and $b \geq 0$ if n is even or

for any real number a if n is odd.

Note that when n is an even integer, $\sqrt[n]{a}$ is defined for nonnegative values of a.

Properties of Nth Roots

a and b are real numbers. m and n are positive integers. Each property is valid for all values of a and b for which the equation is defined.

▶ $\left(\sqrt[n]{a}\right)^n = a$

▶ $\sqrt[n]{a^n} = |a|$ if n is even or

$\qquad = a$ if n is odd

▶ $\sqrt[n]{ab} = \sqrt[n]{a}\,\sqrt[n]{b}$

▶ $\sqrt[n]{\dfrac{a}{b}} = \dfrac{\sqrt[n]{a}}{\sqrt[n]{b}} \qquad b \neq 0$

▶ $\sqrt[m]{\sqrt[n]{a}} = \sqrt[mn]{a} = \sqrt[n]{\sqrt[m]{a}}$

Powers and Roots

The following table shows the squares, square roots, cubes, and cubic roots of the numbers from 1 to 50. Where necessary, the values have been rounded to the nearest thousandth.

n	n^2	$\sqrt{n}$	n^3	$\sqrt[3]{n}$
1	1	1.000	1	1.000
2	4	1.414	8	1.260
3	9	1.732	27	1.442
4	16	2.000	64	1.587
5	25	2.236	125	1.710
6	36	2.449	216	1.817
7	49	2.646	343	1.913
8	64	2.828	512	2.000
9	81	3.000	729	2.080
10	100	3.162	1,000	2.154
11	121	3.317	1,331	2.224
12	144	3.464	1,728	2.289
13	169	3.606	2,197	2.351
14	196	3.742	2,744	2.410
15	225	3.873	3,375	2.466
16	256	4.000	4,096	2.520
17	289	4.123	4,913	2.571
18	324	4.243	5,832	2.621
19	361	4.359	6,859	2.668
20	400	4.472	8,000	2.714
21	441	4.583	9,261	2.759
22	484	4.690	10,648	2.802
23	529	4.796	12,167	2.844
24	576	4.899	13,824	2.884
25	625	5.000	15,625	2.924
26	676	5.099	17,576	2.962
27	729	5.196	19,683	3.000
28	784	5.292	21,952	3.037
29	841	5.385	24,389	3.072
30	900	5.477	27,000	3.107
31	961	5.568	29,791	3.141
32	1,024	5.657	32,768	3.175
33	1,089	5.745	35,937	3.208
34	1,156	5.831	39,304	3.240
35	1,225	5.916	42,875	3.271
36	1,296	6.000	46,656	3.302
37	1,369	6.083	50,653	3.332
38	1,444	6.164	54,872	3.362
39	1,521	6.245	59,319	3.391
40	1,600	6.325	64,000	3.420
41	1,681	6.403	68,921	3.448
42	1,764	6.481	74,088	3.476
43	1,849	6.557	79,507	3.503
44	1,936	6.633	85,184	3.530
45	2,025	6.708	91,125	3.557
46	2,116	6.782	97,336	3.583
47	2,209	6.856	103,823	3.609
48	2,304	6.928	110,592	3.634
49	2,401	7.000	117,649	3.659
50	2,500	7.071	125,000	3.684

LIST **168**

Conditions for Simplifying Radical Expressions

An expression having a square root radical is in its simplest form when:

▶ No radicand has a square factor other than 1.

▶ The radicand is not a fraction.

▶ No radicals are in the denominator.

This can be extended for all radical expressions (not only square root radicals) by adding an additional stipulation.

A radical expression is in simplest form when:

▶ None of the factors of the radicand can be written as powers greater than or equal to the index. (No perfect squares, except 1, can be factors of the quantity under the square root sign; no perfect cubes, except 1, can be factors of the quantity under the cube root sign; and so on.)

▶ The radicand is not a fraction.

▶ No radicals are in the denominator.

Steps for Solving a Quadratic Equation

There are four methods to solving quadratic equations: (1) factoring, (2) using the square root property, (3) completing the square, and (4) using the quadratic formula. Factoring and using the square root property may be used only under certain conditions; completing the square and using the quadratic formula can always be used. A description of each method and when it may be best used follows.

Any quadratic equation has at most two solutions. Some may have the same solution twice. Others may have no real solutions. Some may have only one solution.

Factoring

Use this method if $ax^2 + bx + c$ can be factored.

1. Write the equation in the form $ax^2 + bx + c$. a, b, and c are real numbers, $a \neq 0$.

2. Factor the polynomial.

3. Use the Zero-Product Property to set each factor equal to zero.

4. Solve each linear equation that results.

Solve $x^2 + 3x = 4$.
$x^2 + 3x - 4 = 0$

$(x + 4)(x - 1) = 0$

$(x + 4) = 0 \qquad (x - 1) = 0$
$ x = -4 \qquad x = 1$

Using the Square Root Property

Use this method if $x^2 = k$ or $(ax + b)^2 = k$, $k \neq 0$.

1. Transform the equation so that a perfect square is on one side of the equation and a constant greater than or equal to zero is on the other.

2. Use the Square Root Property to find the square root of each number. Remember that finding the square root of a constant yields positive and negative values.

3. Solve each resulting equation. (If you are finding the square root of a negative number, there is no real solution.)

Solve $x^2 - 16 = 0$.
$x^2 = 16$

$\sqrt{x^2} = \pm\sqrt{16}$

$x = \pm 4$

LIST 169

(Continued)

Completing the Square

This method may always be used to solve quadratic equations. It is best used, however, if the coefficient of the linear term is even.

1. Transform the equation so that the quadratic term plus the linear term equals a constant.

   Solve $x^2 + 12x + 2 = 0$.
   $x^2 + 12x = -2$

2. Divide each term by the coefficient of the quadratic term if it does not equal 1.

3. Complete the square:
   - Multiply the coefficient of x by $\frac{1}{2}$.

     $12 \cdot \frac{1}{2} = 6$

   - Square this value.

     $6^2 = 36$

   - Add the result to both sides of the equation.

     $x^2 + 12x + 36 = -2 + 36$

   - Express one side of the equation as the square of a binomial and the other as a constant.

     $(x + 6)^2 = 34$

4. Follow steps 2 and 3 of Using the Square Root Property.

   $\sqrt{(x+6)^2} = \pm 34$
   $x + 6 = \pm 34$
   $x = -6 + \sqrt{34}$
   $x = -6 - \sqrt{34}$

Using the Quadratic Formula

This method may always be used for any equation of the form $ax^2 + bx + c = 0$.

1. Write the equation in the form $ax^2 + bx + c = 0$. a, b, and c are real numbers, $a \neq 0$.

   Solve $3x^2 = 8x - 2$
   $3x^2 - 8x + 2 = 0$

2. The two roots (if they exist) are $x = \dfrac{-b \pm \sqrt{b^2 - 4ac}}{2a}$.

   $x = \dfrac{8 \pm \sqrt{64 - 24}}{6}$

   $x = \dfrac{8 \pm \sqrt{40}}{6}$

   $x = \dfrac{8 \pm 2\sqrt{10}}{6}$

   $x = \dfrac{4 \pm \sqrt{10}}{3}$

LIST **170**

Imaginary Numbers and Their Powers

Equations such as $x^2 = -25$ have no solution in the set of real numbers. Equations like this have solutions that fall in the realm of imaginary numbers. Imaginary numbers use the imaginary unit i, which is defined as the square root of -1.

$$i = \sqrt{-1} \qquad i^2 = -1$$

The solution to the equation $x^2 = -25$ in the set of imaginary numbers is $x = \pm 5i$. Generally, if $r > 0$, then $\sqrt{-r} = i\sqrt{r}$.

Some interesting patterns emerge for the powers of i as the exponent increases, as shown below.

$i = \sqrt{-1}$ (by definition)

$i^2 = -1$ (by definition)

$i^3 = i^2 \cdot i = -1 \cdot i = -i$

$i^4 = i^2 \cdot i^2 = -1 \cdot -1 = 1$

$i^5 = i^4 \cdot i = 1 \cdot i = i$

$i^6 = i^3 \cdot i^3 = -i \cdot -i = -1$

$i^7 = i^3 \cdot i^4 = -i \cdot 1 = -i$

$\vdots$

Any positive integer power of i equals 1, -1, or i or $-i$.

LIST 171

Discriminant and Coefficients

A quadratic equation of the form $ax^2 + bx + c = 0$ where a, b, and c are real numbers, $a \neq 0$, has a discriminant equal to $b^2 - 4ac$, which determines the number and the kinds of solutions to the equation.

If $b^2 - 4ac$ is	then the equation will have
negative	two (conjugate) imaginary numbers
zero	one real root (double real root)
positive	two different real roots

For any rational numbers a, b, and c, $a \neq 0$:

If $b^2 - 4ac$ is	then the two roots are
positive and the square of a rational number	rational
positive and is not the square of a rational number	irrational

For any quadratic equation of the form $ax^2 + bx + c = 0$, and a, b, and c are real numbers, $a \neq 0$, the sum of the roots equals $-\dfrac{b}{a}$ and the product of the roots equals $\dfrac{c}{a}$.

LIST 172

Quadratic Functions

The function f given by $f(x) = ax^2 + bx + c$ is a quadratic function, provided $a \neq 0$. If its domain is all the real numbers, then:

► Its graph is a parabola.

► Its vertex is the highest or lowest point on the graph, depending on the value of a.

► If $a > 0$, the vertex is the lowest point and the parabola opens upward.

► If $a < 0$, the vertex is the highest point and the parabola opens downward.

► The vertex is the point $\left(-\dfrac{b}{2a}, c - \dfrac{b^2}{4a} \right)$.

► The axis of symmetry is the line $x = -\dfrac{b}{2a}$.

► The x-intercepts (if any) can be found by solving $f(x) = 0$ for x by factoring or by using the quadratic formula.

► The y-intercept is c.

For reference, the quadratic formula is stated below:

If $ax^2 + bx + c = 0$, $a \neq 0$, and $b^2 - 4ac \geq 0$, then $x = \dfrac{-b \pm \sqrt{b^2 - 4ac}}{2a}$.

LIST **173**

Graph of a Circle

A *circle* is one of the conic sections consisting of all points in a plane at a fixed distance from a fixed point.

Some important properties of the graph of a circle centered at the origin are shown below:

- ► The standard form is $x^2 + y^2 = r^2$. $r \geq 0$

  (If $r = 0$, then the circle is called a point-circle.)
- ► The center is $(0,0)$.
- ► The extreme points are $(-r,0)$, $(r,0)$, $(0,r)$, and $(0, -r)$.
- ► Lines of symmetry are infinite in number and include $x = 0$ and $y = 0$.
- ► The x-intercepts are r and $-r$.
- ► The y-intercepts are r and $-r$.
- ► This circle is not a function.

The properties of the graph of a circle centered at (h,k) include:

- ► The standard form is $(x - h)^2 + (y - k)^2 = r^2$. $r \geq 0$

  (If $r = 0$, then the circle is called a point-circle.)
- ► The center is (h,k).
- ► The extreme points are $(h - r, k)$, $(h + r, k)$, $(h, k + r)$, and $(h, k - r)$.
- ► Lines of symmetry are infinite in number and include $x = h$ and $y = k$.
- ► Set $y = 0$ and solve for x to find the x-intercepts (if they exist).
- ► Set $x = 0$ and solve for y to find the y-intercepts (if they exist).
- ► The graph is not a function.

LIST 174

Graph of an Ellipse

An *ellipse* is an elongated circle. It is one of the four conic sections.

The properties of the graph of an ellipse centered at the origin are provided below:

▶ The standard form is $\frac{x^2}{a^2} + \frac{y^2}{b^2} = 1$; $a > 0$, $b > 0$.

▶ The center is $(0,0)$.

▶ The extreme points are $(a,0)$, $(-a,0)$, $(0,b)$, and $(0, -b)$.

▶ The lines of symmetry are $x = 0$ and $y = 0$.

▶ The x-intercepts are a and $-a$.

▶ The y-intercepts are b and $-b$.

▶ This ellipse is not a function.

▶ If $a = b$, then the ellipse is a circle.

▶ If $a > b$, then the x-axis is the major axis.

▶ If $a < b$, then the y-axis is the major axis.

The properties of the graph of an ellipse centered at (h,k) include:

▶ The standard form is $\frac{(x-h)^2}{a^2} + \frac{(y-k)^2}{b^2} = 1.$ $a > 0$, $b > 0$.

▶ The center is (h,k).

▶ The extreme points are $(h + a,k)$, $(h - a,k)$, $(h,k + b)$, and $(h,k - b)$.

▶ The lines of symmetry are $x = h$ and $y = k$.

▶ Set $y = 0$ and solve for x to find the x-intercepts (if they exist).

▶ Set $x = 0$ and solve for y to find the y-intercepts (if they exist).

▶ This ellipse is not a function.

▶ If $a = b$, then the ellipse is a circle.

▶ If $a > b$, then the line $y = k$ is the major axis.

▶ If $a < b$, then the line $x = h$ is the major axis.

LIST 175

Graph of a Parabola

A *parabola* is a conic section shaped like a fountain. It can open up, down, to the left, or to the right. If it opens up or down, it is a quadratic function. Four general forms are addressed below.

The properties of the graph of a parabola with its vertex at the origin and which opens up or down are as follows:

▶ The standard form is $y = ax^2$. $a \neq 0$.

▶ The vertex is $(0,0)$.

▶ The line of symmetry is $x = 0$.

▶ The y-intercept is 0.

▶ The parabola opens up if $a > 0$.

▶ The parabola opens down if $a < 0$.

▶ This parabola is a function.

The properties of the graph of a parabola with its vertex not at the origin and which opens up or down are as follows:

▶ The standard form is $y = ax^2 + bx + c$. $a \neq 0$.

    This can be transformed to $y = a\left(x + \dfrac{b}{2a}\right)^2 + c - \dfrac{b^2}{4a}$ by completing the square.

▶ The vertex is $\left(-\dfrac{b}{2a}, c - \dfrac{b^2}{4a}\right)$.

▶ The line of symmetry is $x = -\dfrac{b}{2a}$.

▶ Use standard form and set $y = 0$ to find the x-intercept (if it exists).

▶ The y-intercept is c (from the standard form).

▶ The parabola opens up if $a > 0$.

▶ The parabola opens down if $a < 0$.

▶ This parabola is a function.

LIST 175

(Continued)

The properties of the graph of a parabola with its vertex at the origin and which opens left or right are as follows:

▶ The standard form is $x = ay^2$. $a \neq 0$.
▶ The vertex is (0,0).
▶ The line of symmetry is $y = 0$.
▶ The x-intercept is 0.
▶ The parabola opens to the right if $a > 0$.
▶ The parabola opens to the left if $a < 0$.
▶ This parabola is not a function.

The properties of the graph of a parabola with its vertex not at the origin and which opens left or right are as follows:

▶ The standard form is $x = ay^2 + by + c$. $a \neq 0$.

This can be transformed to $x = a\left(y + \dfrac{b}{2a}\right)^2 + c - \dfrac{b^2}{4a}$ by completing the square.

▶ The vertex is $\left(c - \dfrac{b^2}{4a}, \dfrac{b}{2a}\right)$.

▶ The line of symmetry is $y = -\dfrac{b}{2a}$.

▶ The x-intercept is c (from the standard form).
▶ Use the standard form and set $x = 0$ to find the y-intercept (if it exists).
▶ The parabola opens to the right if $a > 0$.
▶ The parabola opens to the left if $a < 0$.
▶ This parabola is not a function.

LIST 176

Graph of a Hyperbola

A *hyperbola,* the fourth and final conic section, is a curve with two branches, each of which approaches other lines called asymptotes. The asymptotes are not part of the graph, but they are helpful in drawing the graph. A hyperbola may open to the right and left or up and down.

The properties of the graph of a hyperbola with the center at the origin and which opens right and left are as follows:

▶ The standard form is $\dfrac{x^2}{a^2} - \dfrac{y^2}{b^2} = 1.$ $a \neq 0, \ b \neq 0.$

▶ The center is $(0,0)$.

▶ The vertices are $(-a,0)$ and $(a,0)$.

▶ The lines of symmetry are $x = 0$ and $y = 0$.

▶ The asymptotes are $y = -\dfrac{b}{a}x$ and $y = \dfrac{b}{a}x$.

▶ The x-intercepts are $-a$ and a.

▶ The y-intercepts do not exist.

▶ The hyperbola opens to the right and left.

▶ This hyperbola is not a function.

The properties of the graph of a hyperbola that opens right and left and is centered at a point other than the origin are as follows:

▶ The standard form is $\dfrac{(x-h)^2}{a^2} - \dfrac{(y-k)^2}{b^2} = 1.$ $a \neq 0, \ b \neq 0.$

▶ The center is (h,k).

▶ The vertices are $(h-a,k)$ and $(h+a,k)$.

▶ The lines of symmetry are $x = h$ and $y = k$.

▶ The asymptotes are $y - k = \dfrac{b}{a}(x-h)$ and $y - k = -\dfrac{b}{a}(x-h)$.

▶ The hyperbola opens to the right and left.

▶ This hyperbola is not a function.

LIST 176

(Continued)

The properties of the graph of a hyperbola that opens up and down and is centered at the origin are as follows:

▶ The standard form is $\dfrac{y^2}{b^2} - \dfrac{x^2}{a^2} = 1.$ $a \neq 0,\ b \neq 0.$

▶ The center is $(0,0)$.

▶ The vertices are $(0,b)$ and $(0,-b)$.

▶ The lines of symmetry are $x = 0$ and $y = 0$.

▶ The asymptotes are $y = -\dfrac{b}{a}x$ and $y = \dfrac{b}{a}x.$

▶ The x-intercepts do not exist.

▶ The y-intercepts are b and $-b$.

▶ The hyperbola opens up and down.

▶ This hyperbola is not a function.

The properties of the graph of a hyperbola that opens up and down and is centered at a point other than the origin are as follows:

▶ The standard form is $\dfrac{(y-k)^2}{b^2} - \dfrac{(x-h)^2}{a^2} = 1.$ $a \neq 0,\ b \neq 0.$

▶ The center is (h,k).

▶ The vertices are $(h, k + b)$ and $(h, k - b)$.

▶ The lines of symmetry are $x = h$ and $y = k$.

▶ The asymptotes are $y - k = \dfrac{b}{a}(x - h)$ and $y - k = -\dfrac{b}{a}(x - h).$

▶ The hyperbola opens up and down.

▶ This hyperbola is not a function.

Properties of Complex Numbers

A *complex number* is any number of the form $a + bi$ where a and b are real numbers and $i^2 = -1$. The set of all complex numbers $a + bi$ with $b = 0$ is the set of real numbers. In a complex number of the form $a + bi$, a is called the *real* part and b is called the *imaginary* part. If $a = 0$, then the complex number of the form $a + bi$ is called *pure imaginary.*

Equality is defined as follows:

$a + bi = c + di$ if and only if $a = c$ and $b = d$.

Many of the properties of real numbers are also properties of complex numbers. In the properties of complex numbers listed below, w, y, and z are complex numbers.

Closure	$w + y$ is a unique complex number.
	$w \cdot y$ is a unique complex number.
Commutative Laws	$w + y = y + w$
	$w \cdot y = y \cdot w$
Associative Laws	$w + (y + z) = (w + y) + z$
	$w(yz) = (wy)z$
Identity Laws	$w + 0 = w$
	$w \cdot 1 = w$
Distributive Law	$w(y + z) = wy + wz$
Additive Inverse	$w + (-w) = 0$
Multiplicative Inverse	$w \cdot w^{-1} = 1 \qquad w \neq 0$

LIST 178

Operations with Complex Numbers

Operations with complex numbers can be compared to corresponding operations for polynomials. In the operations below, $a + bi$ and $c + di$ are complex numbers. c is any real number.

Addition	Add the real parts and add the imaginary parts. $(a + bi) + (c + di) = (a + c) + (b + d)i$
Subtraction	Subtract the real parts and subtract the imaginary parts. $(a + bi) - (c + di) = (a - c) + (b - d)i$
Distributivity	Multiply the real part by c and multiply the imaginary part by c. $c(a + bi) = ca + cbi$
Multiplication	Carry out the multiplication as if the numbers were binomials and replace i^2 with -1. $(a + bi)(c + di) = ac + adi + bci + bdi^2 = (ac - bd) + (ad + bc)i$
Division	Multiply both the numerator and denominator of the fraction by the conjugate of the denominator. See List 179, "Conjugate Complex Numbers." $\dfrac{a + bi}{c + di} = \dfrac{a + bi}{c + di} \cdot \dfrac{\overline{c + di}}{c + di} = \dfrac{ac + bd}{c^2 + d^2} + \dfrac{bc - ad}{c^2 + d^2}\,i$ $c + di \neq 0 + 0i$

LIST 179

Conjugate Complex Numbers

Conjugate complex numbers, which are also called *complex imaginaries,* are complex numbers that are identical except that the pure imaginary terms have opposite signs or are both zero.

$a + bi$ and $a - bi$ are conjugate complex numbers, where a and b are real numbers and $i^2 = -1$, and bi and $-bi$ are pure imaginary terms.

The conjugate of a complex number is denoted by a raised line $^-$. $\overline{a + bi}$ is read "the conjugate of $a + bi$" and equals $a - bi$.

Important properties of conjugates follow:

▶ $\overline{a} = a$ where a is a real number because $a = a + 0i$ and $a = a - 0i = a$.

▶ $z \cdot \overline{z} = (a + bi)(a - bi) = a^2 - b^2$ where $z = a + bi$.

▶ $\overline{z + w} = \overline{z} + \overline{w}$. The conjugate of a sum is the sum of the conjugates. (z and w represent complex numbers.)

▶ $\overline{zw} = \overline{z} \cdot \overline{w}$. The conjugate of a product is the product of the conjugates. (z and w represent complex numbers.)

▶ $\overline{z^n} = (\overline{z})^n$ where z is a complex number and n is a positive integer.

▶ $\overline{\left(\dfrac{z}{w}\right)} = \dfrac{\overline{z}}{\overline{w}}$ where z and w represent complex numbers, $w \neq 0 + 0i$.

LIST 180

Vectors

A *vector* describes quantities that involve both magnitude (size) and direction. In geometry, a vector is a directed line segment, often described by a line segment with an arrow at one end. In algebra, a vector is described by the coordinates of the initial and terminal points of the directed line segments.

Definitions and notations about vectors are listed below.

▶ The ordered pair notation to describe a vector is $\langle a_1, a_2 \rangle$.

▶ The components of a vector are the numbers a_1 and a_2.

▶ $\overrightarrow{PQ}$ is the vector associated with the directed line segment with initial point $P = (x_0, y_0)$ and terminal point $Q = (x_1, y_1)$. This vector has components $x_1 - x_0$ and $y_1 - y_0$.

▶ Vectors such as $\langle a_1, a_2 \rangle$ and $\langle b_1, b_2 \rangle$ are equal if $a_1 = b_1$ and $a_2 = b_2$.

▶ Vectors are denoted by lowercase boldface letters, starting at the beginning of the alphabet. These letters are used to distinguish vectors from numbers that are sometimes called *scalars*. Since it is difficult to write a boldface letter by hand, a vector may be written as a lowercase letter with an arrow over it.

▶ A zero vector, $\langle 0, 0 \rangle$, is noted by **0**.

▶ The length (or norm) of a vector is $|a| = \sqrt{a_1^2 + a_2^2}$.

▶ A unit vector has a length of 1. $\boldsymbol{i} = \langle 1, 0 \rangle$ and $\boldsymbol{j} = \langle 0, 1 \rangle$ are unit vectors.

▶ A combination of vectors is $\boldsymbol{a} = \langle a_1, a_2 \rangle$ and $\boldsymbol{b} = \langle b_1, b_2 \rangle$.
Other examples include:

$$\boldsymbol{a} + \boldsymbol{b} = \langle a_1 + b_1, a_2 + b_2 \rangle$$
$$\boldsymbol{a} - \boldsymbol{b} = \langle a_1 - b_1, a_2 - b_2 \rangle$$
$$c\boldsymbol{a} = \langle ca_1, ca_2 \rangle \text{ where } c \text{ is a number}$$

LIST 180

(Continued)

▶ Properties of vectors include:

$$0 + a = a + 0 = a$$
$$0a = 0$$
$$c0 = 0$$
$$a + (b + c) = (a + b) + c$$
$$a - b = a + (-1)b$$
$$a + (-a) = 0$$
$$a + b = b + a$$
$$1a = a$$
$$c(a + b) = ca + cb$$
$$|ca| = |c||a|$$

Any vector can be expressed $a = \langle a_1, a_2 \rangle$ as a combination of unit vectors i and j. Note the following:

$$a = a_1 i + a_2 j$$
$$|a_1 i + a_2 j| = \sqrt{a_1^2 + a_2^2}$$
$$(a_1 i + a_2 j) + (b_1 i + b_2 j) = (a_1 + b_1)i + (a_2 + b_2)j$$
$$(a_1 i + a_2 j) - (b_1 i + b_2 j) = (a_1 - b_1)i + (a_2 - b_2)j$$
$$c(a_1 i + a_2 j) = ca_1 i + ca_2 j$$

Matrices

A *matrix* is a set of quantities arranged in rows and columns to form a rectangular array, usually enclosed in parentheses. Matrices do not have a numerical value; they are used to represent relations between quantities. Matrices may also be used to represent and solve simultaneous equations.

Facts about matrices are shown below.

▶ If there are m rows and n columns, the matrix is an $m \times n$ matrix. This is called the order of the matrix.

▶ Matrices are named with capital letters.

▶ Individual members in a matrix are called elements (or entries) of the matrix.

▶ Particular elements may be identified by the horizontal row and vertical column to which they belong. a_{ij} denotes the element in the ith row and jth column of matrix A.

▶ Two matrices are equal if they are the same size and if all the corresponding elements are the same.

▶ To add matrices, add corresponding elements together to obtain another matrix of the same order.

▶ Only matrices of the same order may be added.

▶ To subtract matrices, subtract corresponding elements to obtain another matrix of the same order.

▶ Only matrices of the same order may be subtracted.

▶ To multiply a matrix by a number (also called a scalar), multiply each element by the scalar.

▶ To multiply matrices, multiply row by column and add.

For example:

$$\begin{bmatrix} 1 & 2 & 3 \\ 4 & 5 & 6 \end{bmatrix} \times \begin{bmatrix} 8 & 11 & 14 \\ 9 & 12 & 15 \\ 10 & 13 & 16 \end{bmatrix} = \begin{bmatrix} 56 & 72 & 92 \\ 137 & 182 & 227 \end{bmatrix}$$

$$1 \times 8 + 2 \times 9 + 3 \times 10 = 56$$
$$1 \times 11 + 2 \times 12 + 3 \times 13 = 72$$
$$1 \times 14 + 2 \times 15 + 3 \times 16 = 92$$
$$4 \times 8 + 5 \times 9 + 6 \times 10 = 137$$
$$4 \times 11 + 5 \times 12 + 6 \times 13 = 182$$
$$4 \times 14 + 5 \times 15 + 6 \times 16 = 227$$

LIST 181

(Continued)

▶ Two matrices *A* and *B* may be multiplied only if the number of columns in *A* is the same as the number of rows in *B*. Multiplication is not commutative.

▶ The determinant is a function of a square matrix derived by multiplying and adding the elements together to obtain a single number.

  ▪ The determinant of a 1×1 matrix $[a_1]$ is its element.

  ▪ The determinant of a 2×2 matrix $\begin{bmatrix} a_1 & b_1 \\ a_2 & b_2 \end{bmatrix} = a_1 b_2 - a_2 b_1$

  ▪ The determinant of a 3×3 matrix $\begin{bmatrix} a_1 & b_1 & c_1 \\ a_2 & b_2 & c_2 \\ a_3 & b_3 & c_3 \end{bmatrix} =$

$$a_1 b_2 c_3 + a_2 b_3 c_1 + a_3 b_1 c_2 - a_1 b_3 c_2 - a_2 b_1 c_3 - a_3 b_2 c_1$$

▶ For any square matrix *A*, *A* has an inverse denoted A^{-1} if and only if the determinant of *A* does not equal zero.

Types of Matrices

Several specific types of matrices and their distinguishing characteristics are noted below.

Square Matrix—a matrix that has the same number of rows and columns. The diagonal from the top left to the bottom right is the leading diagonal (or principal diagonal). The sum of the elements in this diagonal is called the trace, or spur, of the matrix.

Row Matrix—a matrix with only one row.

Column Matrix—a matrix with only one column.

Zero Matrix (*Null Matrix*)—a matrix in which all the elements are equal to zero.

Unit Matrix (*Identity Matrix*)—a square matrix in which all the elements in the leading diagonal are one and the other elements are equal to zero.

Diagonal Matrix—a square matrix in which all the elements are zero except those in the leading diagonal.

Triangular Matrix—a square matrix in which either all the elements above the leading diagonal are zero or all the elements below the leading diagonal are zero.

Conformable Matrices—two matrices in which the number of columns in one is the same as the number of rows in the other.

Transpose of a Matrix—the matrix that results from interchanging the rows and columns.

Negative of a Matrix—the matrix whose elements are opposite each corresponding element of the original matrix.

Trigonometry and Calculus

LIST 183

Right Triangle Definitions of Trigonometric Functions

The six functions that apply to right triangles are sine, cosine, tangent, cotangent, secant, and cosecant. The following formulas will help you calculate the values of these trigonometric functions.

$\triangle ABC$ is a right triangle. $\angle C$ is a right angle. c is the hypotenuse, b is the adjacent side with respect to $\angle A$, and a is the opposite side with respect to $\angle A$.

sine A = sin A
$$= \frac{\text{length of the opposite side}}{\text{length of the hypotenuse}} = \frac{a}{c}$$

cosine A = cos A
$$= \frac{\text{length of the adjacent side}}{\text{length of the hypotenuse}} = \frac{b}{c}$$

tangent A = tan A
$$= \frac{\text{length of the opposite side}}{\text{length of the adjacent side}} = \frac{a}{b}$$

cotangent A = cot A
$$= \frac{\text{length of the adjacent side}}{\text{length of the opposite side}} = \frac{b}{a}$$

secant A = sec A
$$= \frac{\text{length of the hypotenuse}}{\text{length of the adjacent side}} = \frac{c}{b}$$

cosecant A = csc A
$$= \frac{\text{length of the hypotenuse}}{\text{length of the opposite side}} = \frac{c}{a}$$

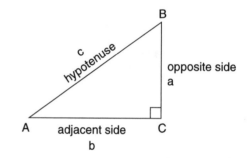

Sin A, cos A, tan A, cot A, sec A, and csc A do not depend on the lengths of the sides of the triangle but only on the ratios of the lengths of the two corresponding sides.

Each of the trigonometric values is nonnegative.

LIST **184**

Trigonometric Functions of Complementary Angles

The acute angles of any right triangle are complementary. In the example below, $\angle A$ and $\angle B$ are complementary angles because the $m\angle A + m\angle B = 90°$.

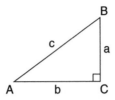

The following formulas associate functions in pairs: sine and cosine, tangent and cotangent, and secant and cosecant.

$$\sin A = \frac{a}{c} = \cos B$$

$$\cos A = \frac{b}{c} = \sin B$$

$$\tan A = \frac{a}{b} = \cot B$$

$$\cot A = \frac{b}{a} = \tan B$$

$$\sec A = \frac{c}{b} = \csc B$$

$$\csc A = \frac{c}{a} = \sec B$$

Each function of the pair is called a *cofunction* of the other. Any function of an acute angle is equal to the corresponding cofunction of the complement of the angle.

LIST **185**

How to Express Angles in Circular Measure

The familiar units used to measure angles are degrees, minutes, and seconds. One complete circular revolution is partitioned into 360 equal units called *degrees*. Each degree is subdivided into 60 equal units called *minutes*. Each minute is further subdivided into 60 equal units called *seconds*. ° is the symbol for degree, ' is the symbol for minutes, and " is the symbol for seconds.

$1° = 60'$ and $1' = 60''$

$39°14'28''$ is read "39 degrees, 14 minutes, and 28 seconds."

To change from degrees, minutes, and seconds to decimal notation, multiply the number of degrees by 1, the number of minutes by $\frac{1}{60}$, and the number of seconds by $\frac{1}{3600}$, for example, $39°14'28'' = 39.24\bar{1}°$.

An alternate system of measuring angles is the use of a unit called a radian. A *radian* (rad) is the measure of the central angle subtended by an arc of a circle whose length equals the radius of the circle. Any circle contains a total of 2π radians. A straight angle, which cuts off a semicircle, has a measure of π radians.

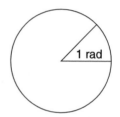

$1 \text{ radian} = \frac{180}{\pi} \text{ degrees} \approx 57°17'45'' \approx 57.296°$

$1 \text{ degree} = \frac{\pi}{180} \text{ radians} \approx 0.01745 \text{ rad}$

Conversions

▶ To convert radians to degrees, multiply the number of radians by $\frac{180}{\pi}$.

▶ To convert degrees to radians, multiply the number of degrees by $\frac{\pi}{180}$.

It is customary to omit the unit "radian." Therefore we write $1° = \frac{\pi}{180}$. The radian measures of some common angles follow.

Some Important Equivalents

$0° = 0$	$90° = \frac{\pi}{2}$	$180° = \pi$
$15° = \frac{\pi}{12}$	$105° = \frac{7\pi}{12}$	$225° = \frac{5\pi}{4}$
$30° = \frac{\pi}{6}$	$120° = \frac{2\pi}{3}$	$270° = \frac{3\pi}{2}$
$45° = \frac{\pi}{4}$	$135° = \frac{3\pi}{4}$	$315° = \frac{7\pi}{4}$
$60° = \frac{\pi}{3}$	$150° = \frac{5\pi}{6}$	$360° = 2\pi$
$75° = \frac{5\pi}{12}$	$165° = \frac{11\pi}{12}$	

LIST **186**

Circular Function Definition
of Trigonometric Functions

The *circular definition of trigonometric functions* is used in physics, electronics, and biology where the periodic nature of the functions is emphasized. The circular function definition allows angles with measures greater than 90°, greater than 360°, or less than 0°. These angles are not defined by the right triangle definition.

In the circular definition, θ is any angle in *standard position.* This means that the angle is on the coordinate plane in such a way that the vertex of the angle is aligned with the origin. Its *initial side* coincides with the x-axis. The other side of the angle is called the *terminal side.* The direction that indicates rotation from the initial side of the angle to the terminal side of the angle is shown by the arrow. (x,y) is a point on the terminal ray of the angle.

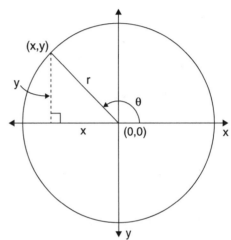

$$r = \sqrt{x^2 + y^2}$$

$$\sin \theta = \frac{y}{r} \qquad \csc \theta = \frac{r}{y}$$

$$\cos \theta = \frac{x}{r} \qquad \sec \theta = \frac{r}{x}$$

$$\tan \theta = \frac{y}{x} \qquad \cot \theta = \frac{x}{y}$$

The following trigonometric reciprocals result from the definitions:

$$\sin \theta = \frac{1}{\csc \theta} \qquad \csc \theta = \frac{1}{\sin \theta}$$

$$\cos \theta = \frac{1}{\sec \theta} \qquad \sec \theta = \frac{1}{\cos \theta}$$

$$\tan \theta = \frac{1}{\cot \theta} \qquad \cot \theta = \frac{1}{\tan \theta}$$

$$\tan \theta = \frac{\sin \theta}{\cos \theta} \qquad \cot \theta = \frac{\cos \theta}{\sin \theta}$$

LIST 187

Trigonometric Functions of Quadrantal Angles

Angles may be measured in radians or degrees. *Quadrantal angles* are angles whose measures are:

- ▶ 0° or 0 radians

- ▶ 90° or $\frac{\pi}{2}$ radians

- ▶ 180° or π radians

- ▶ 270° or $\frac{3\pi}{2}$ radians

- ▶ 360° or 2π radians

- ▶ and any angles coterminal with them.

Coterminal angles are two angles that, when placed in standard form, have the same terminal side.

- ▶ To find coterminal angles of angles measured in degrees, add multiples of 360° to the measure of a given angle.

- ▶ To find coterminal angles of angles measured in radians, add multiples of 2π to the measure of the given angle.

The following values of the quadrantal angles were obtained from the definitions of the trigonometric functions. In some cases, substitutions will result in a zero in the denominator and the functions are undefined for these values. Some texts use ∞ or ±∞ to indicate that the function is undefined.

In Radians	*In Degrees*	*Sin*	*Cos*	*Tan*	*Cot*	*Sec*	*Csc*
0	0°	0	1	0	∞	1	∞
$\frac{\pi}{2}$	90°	1	0	∞	0	∞	1
π	180°	0	−1	0	∞	−1	∞
$\frac{3\pi}{2}$	270°	−1	0	∞	0	∞	−1

- ▶ Sine and cosine are defined for all values.

- ▶ Remaining functions are not defined when their denominators are 0.

LIST **188**

The Quadrant Signs of the Functions

In the circular function definition of the trigonometric functions, *r*, the radius of the circle, is always positive. The quadrant signs of *x* and *y* determine the signs of the various trigonometric functions.

An angle in the standard position that terminates in the first quadrant is called a *first-quadrant angle* or an *angle in the first quadrant*.

An angle in the standard position that terminates in the second quadrant is called a *second-quadrant angle* or an *angle in the second quadrant*.

The definitions are similar for the third and fourth quadrants.

The chart below summarizes the signs of the trigonometric functions in each quadrant.

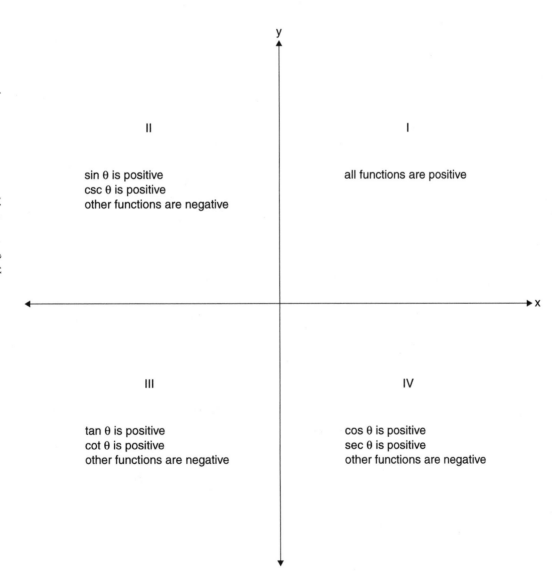

LIST **189**

Variations of Trigonometric Functions

Each trigonometric function varies as θ increases continuously from 0° to 360°. The following chart shows how each of the trigonometric functions varies.

As θ increases from	0° to 90° or 0 to $\frac{\pi}{2}$ rad	90° to 180° or $\frac{\pi}{2}$ rad to π rad	180° to 270° or π rad to $\frac{3\pi}{2}$ rad	270° to 360° or $\frac{3\pi}{2}$ rad to 2π rad
$\sin \theta$	increases from 0 to 1	decreases from 1 to 0	decreases from 0 to −1	increases from −1 to 0
$\cos \theta$	decreases from 1 to 0	decreases from 0 to −1	increases from −1 to 0	increases from 0 to 1
$\tan \theta$	increases from 0 to +∞	increases from −∞ to 0	increases from 0 to +∞	increases from −∞ to 0
$\cot \theta$	decreases from +∞ to 0	decreases from 0 to −∞	decreases from +∞ to 0	decreases from 0 to −∞
$\sec \theta$	increases from 1 to +∞	increases from −∞ to −1	decreases from −1 to −∞	decreases from +∞ to 1
$\csc \theta$	decreases from +∞ to 1	increases from 1 to +∞	increases from −∞ to −1	decreases from −1 to −∞

Summary

► The sine and cosine take on values between −1 and 1 inclusive.

► The tangent and cotangent can take on any values.

► The secant and cosecant can take on any values except those between −1 and 1.

LIST 190
Trigonometric Functions of Some Special Angles

Some special angles have trigonometric function values that can be computed precisely, in terms of radicals. The following table summarizes those values.

angle	sin	cos	tan	cot	sec	csc
$0° = 0$	0	1	0	∞	1	∞
$15° = \dfrac{\pi}{12}$	$\dfrac{\sqrt{2}}{4}\left(\sqrt{3}-1\right)$	$\dfrac{\sqrt{2}}{4}\left(\sqrt{3}+1\right)$	$2-\sqrt{3}$	$2+\sqrt{3}$	$\sqrt{2}\left(\sqrt{3}-1\right)$	$\sqrt{2}\left(\sqrt{3}+1\right)$
$30° = \dfrac{\pi}{6}$	$\dfrac{1}{2}$	$\dfrac{\sqrt{3}}{2}$	$\dfrac{\sqrt{3}}{3}$	$\sqrt{3}$	$\dfrac{2\sqrt{3}}{3}$	2
$45° = \dfrac{\pi}{4}$	$\dfrac{\sqrt{2}}{2}$	$\dfrac{\sqrt{2}}{2}$	1	1	$\sqrt{2}$	$\sqrt{2}$
$60° = \dfrac{\pi}{3}$	$\dfrac{\sqrt{3}}{2}$	$\dfrac{1}{2}$	$\sqrt{3}$	$\dfrac{\sqrt{3}}{3}$	2	$\dfrac{2\sqrt{3}}{3}$
$75° = \dfrac{5\pi}{12}$	$\dfrac{\sqrt{2}}{4}\left(\sqrt{3}+1\right)$	$\dfrac{\sqrt{2}}{4}\left(\sqrt{3}-1\right)$	$2+\sqrt{3}$	$2-\sqrt{3}$	$\sqrt{2}\left(\sqrt{3}+1\right)$	$\sqrt{2}\left(\sqrt{3}-1\right)$
$90° = \dfrac{\pi}{2}$	1	0	∞	0	∞	1
$105° = \dfrac{7\pi}{12}$	$\dfrac{\sqrt{2}}{4}\left(\sqrt{3}+1\right)$	$-\dfrac{\sqrt{2}}{4}\left(\sqrt{3}-1\right)$	$-\left(2+\sqrt{3}\right)$	$-\left(2-\sqrt{3}\right)$	$-\sqrt{2}\left(\sqrt{3}+1\right)$	$\sqrt{2}\left(\sqrt{3}-1\right)$
$120° = \dfrac{2\pi}{3}$	$\dfrac{\sqrt{3}}{2}$	$-\dfrac{1}{2}$	$-\sqrt{3}$	$-\dfrac{\sqrt{3}}{3}$	-2	$\dfrac{2\sqrt{3}}{3}$
$135° = \dfrac{3\pi}{4}$	$\dfrac{\sqrt{2}}{2}$	$-\dfrac{\sqrt{2}}{2}$	-1	-1	$-\sqrt{2}$	$\sqrt{2}$
$150° = \dfrac{5\pi}{6}$	$\dfrac{1}{2}$	$-\dfrac{\sqrt{3}}{2}$	$-\dfrac{\sqrt{3}}{3}$	$-\sqrt{3}$	$-\dfrac{2\sqrt{3}}{3}$	2
$165° = \dfrac{11\pi}{12}$	$\dfrac{\sqrt{2}}{4}\left(\sqrt{3}-1\right)$	$-\dfrac{\sqrt{2}}{4}\left(\sqrt{3}+1\right)$	$-\left(2-\sqrt{3}\right)$	$-\left(2+\sqrt{3}\right)$	$-\sqrt{2}\left(\sqrt{3}-1\right)$	$\sqrt{2}\left(\sqrt{3}+1\right)$
$180° = \pi$	0	-1	0	∞	-1	∞
$\vdots$						
$270° = \dfrac{3\pi}{2}$	-1	0	∞	0	∞	-1
$\vdots$						
$360° = 2\pi$	0	1	0	∞	1	∞

LIST **191**

Fundamental Periods of Trigonometric Functions

Each of the six trigonometric functions is periodic, which means that, geometrically, the graph repeats itself over a period. The *fundamental period* is the smallest possible positive number with this property.

Stated formally: if f is a function defined by the equation $y = f(x)$ and there is a positive number a such that $f(x + a) = f(x)$ for all real numbers for which the function is defined, then f is a periodic function, and a is called the *period of the function*. Furthermore, if a is the smallest positive number with this property, then a is called the *fundamental period of the function*.

A summary of the six trigonometric functions and their fundamental periods follows. x is any real number for which the function is defined. See List 192, "Trigonometric Functions: Their Domains, Ranges, and Periods."

n is an integer.

$\sin x = \sin (x + 2\pi n)$ Fundamental period is 2π.

$\cos x = \cos (x + 2\pi n)$ Fundamental period is 2π.

$\tan x = \tan (x + \pi n)$ Fundamental period is π.

$\cot x = \cot (x + \pi n)$ Fundamental period is π.

$\sec x = \sec (x + 2\pi n)$ Fundamental period is 2π.

$\csc x = \csc (x + 2\pi n)$ Fundamental period is 2π.

Trigonometric Functions: Their Domains, Ranges, and Periods

An important aspect of any function is the values for which it is defined. The domains and ranges of the trigonometric functions, as well as the fundamental periods, are shown below. Restrictions on the domains are also noted. *n* is any integer.

Function	*Domain*	*Range*	*Period*
$y = \sin x$	all real numbers	$-1 \leq y \leq 1$	2π
$y = \cos x$	all real numbers	$-1 \leq y \leq 1$	2π
$y = \tan x$	all real numbers except $\frac{\pi}{2}$ plus integer multiples of π $\left(x \neq \frac{\pi}{2} + n\pi\right)$	all real numbers	π
$y = \cot x$	all real numbers except integer multiples of π $\left(x \neq n\pi\right)$	all real numbers	π
$y = \sec x$	all real numbers except $\frac{\pi}{2}$ plus integer multiples of π $\left(x \neq \frac{\pi}{2} + n\pi\right)$	$\left|y\right| \geq 1$	2π
$y = \csc x$	all real numbers except integer multiples of π $\left(x \neq n\pi\right)$	$\left|y\right| \geq 1$	2π

LIST 193

Graphs of the Trigonometric Functions

The graphs of the trigonometric functions below have interesting features. Note the domain, range, and period of each. See List 191, "Fundamental Periods of Trigonometric Functions," and List 192, "Trigonometric Functions: Their Domains, Ranges, and Periods."

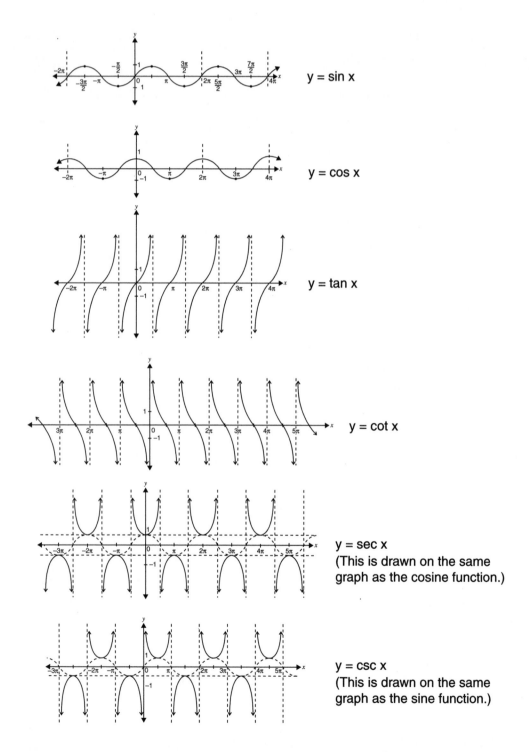

$y = \sin x$

$y = \cos x$

$y = \tan x$

$y = \cot x$

$y = \sec x$
(This is drawn on the same graph as the cosine function.)

$y = \csc x$
(This is drawn on the same graph as the sine function.)

LIST **194**

General Trigonometric Functions: Their Periods and Amplitudes

The *amplitude* of the sine (or cosine) function is defined as half of the difference between its maximum and minimum values. Since the graph of $y = \sin x$ shows $\frac{1}{2}|1 - (-1)| = 1$, the amplitude of the sine (or cosine) function is 1. Understanding the graphs of the basic trigonometric functions will help you to draw conclusions about general trigonometric functions such as:

	Period	Amplitude
$y = a \sin bx$	$\dfrac{2\pi}{|b|}$	$|a|$
$y = a \cos bx$	$\dfrac{2\pi}{|b|}$	$|a|$
$y = a \tan bx$	$\dfrac{\pi}{|b|}$	—
$y = a \cot bx$	$\dfrac{\pi}{|b|}$	—
$y = a \sec bx$	$\dfrac{2\pi}{|b|}$	—
$y = a \csc bx$	$\dfrac{2\pi}{|b|}$	—

Basic Trigonometric Identities

A *trigonometric identity* is an equation that is valid for all the values of the variable for which every expression in the equation is defined. Admissible values are values of the variable for which the trigonometric identity is defined.

    Although there are other trigonometric identities, the ones that follow should be learned in order to simplify trigonometric expressions and prove identities.

Reciprocal Relationships

$$\csc \alpha = \frac{1}{\sin \alpha} \qquad \sin \alpha \neq 0$$

$$\sec \alpha = \frac{1}{\cos \alpha} \qquad \cos \alpha \neq 0$$

$$\cot \alpha = \frac{1}{\tan \alpha} \qquad \tan \alpha \neq 0$$

Quotient Relationships

$$\tan \alpha = \frac{\sin \alpha}{\cos \alpha} \qquad \cos \alpha \neq 0$$

$$\cot \alpha = \frac{\cos \alpha}{\sin \alpha} \qquad \sin \alpha \neq 0$$

Pythagorean Relationships

$$\sin^2 \alpha + \cos^2 \alpha = 1$$

$$1 + \tan^2 \alpha = \sec^2 \alpha$$

$$1 + \cot^2 \alpha = \csc^2 \alpha$$

Odd-Even Identities

$$\sin (-\alpha) = -\sin \alpha$$

$$\cos (-\alpha) = \cos \alpha$$

$$\tan (-\alpha) = -\tan \alpha$$

$$\csc (-\alpha) = -\csc \alpha$$

$$\sec (-\alpha) = \sec \alpha$$

$$\cot (-\alpha) = -\cot \alpha$$

LIST 196

Angle-Sum and Angle-Difference Formulas

Addition formulas for the sine and cosine are formulas that express $\sin(\alpha + B)$ and $\cos(\alpha + B)$ in terms of $\sin\alpha$, $\sin B$, $\cos\alpha$, and $\cos B$. Equivalent forms result from substituting α for B and using $\cos(-B) = \cos B$ and $\sin(-B)$ for $-\sin B$.

$$\sin(\alpha + B) = \sin\alpha\cos B + \cos\alpha\sin B$$

$$\cos(\alpha + B) = \cos\alpha\cos B - \sin\alpha\sin B$$

$$\sin(\alpha - B) = \sin\alpha\cos B - \cos\alpha\sin B$$

$$\cos(\alpha - B) = \cos\alpha\cos B + \sin\alpha\sin B$$

Addition formulas for other functions may be derived from those for sine and cosine. The two most important derivations are the formulas for the tangent.

$$\tan(\alpha + B) = \frac{\tan\alpha + \tan B}{1 - \tan\alpha\tan B} \qquad \tan\alpha\tan B \neq 1$$

$$\tan(\alpha - B) = \frac{\tan\alpha - \tan B}{1 + \tan\alpha\tan B} \qquad \tan\alpha\tan B \neq -1$$

Formulas for the cotangent follow.

$$\cot(\alpha + B) = \frac{\cot B\,\cot\alpha - 1}{\cot B + \cot\alpha} \qquad \cot B \neq -\cot\alpha$$

$$\cot(\alpha - B) = \frac{\cot B\,\cot\alpha + 1}{\cot B - \cot\alpha} \qquad \cot B \neq \cot\alpha$$

LIST **197**

Double-Angle Formulas and Half-Angle Formulas

The following double-angle formulas result when $\alpha = B$ in the sum formulas. The half-angle formulas are derived from the double-angle formulas.

Double-Angle Formulas

$$\sin 2\alpha = 2 \sin\alpha \cos\alpha = \frac{2 \tan\alpha}{1 + \tan^2\alpha}$$

$$\cos 2\alpha = \cos^2\alpha - \sin^2\alpha = 1 - 2\sin^2\alpha = 2\cos^2\alpha - 1 = \frac{1 - \tan^2\alpha}{1 + \tan^2\alpha}$$

$$\tan 2\alpha = \frac{2 \tan\alpha}{1 - \tan^2\alpha}$$

$$\cot 2\alpha = \frac{\cot^2\alpha - 1}{2 \cot\alpha}$$

Half-Angle Formulas

$$\sin \frac{1}{2}\alpha = \sqrt{\frac{1 - \cos\alpha}{2}} \quad \text{if} \quad \sin \frac{1}{2}\alpha \geq 0$$

$$\text{or}$$

$$-\sqrt{\frac{1 - \cos\alpha}{2}} \quad \text{if} \quad \sin \frac{1}{2}\alpha < 0$$

$$\cos \frac{1}{2}\alpha = \sqrt{\frac{1 + \cos\alpha}{2}} \quad \text{if} \quad \cos \frac{1}{2}\alpha \geq 0$$

$$\text{or}$$

$$-\sqrt{\frac{1 + \cos\alpha}{2}} \quad \text{if} \quad \cos \frac{1}{2}\alpha < 0$$

$$\tan \frac{1}{2}\alpha = \frac{\sin\alpha}{1 + \cos\alpha} = \frac{1 - \cos\alpha}{\sin\alpha}$$

$$\cot \frac{1}{2}\alpha = \frac{\sin\alpha}{1 - \cos\alpha} = \frac{1 + \cos\alpha}{\sin\alpha}$$

LIST **198**

Multiple-Angle Formulas, Product Formulas, and Sum and Difference Formulas

The lists below contain formulas for multiple angles (including the general formula), formulas for the product of the sine and cosine functions, and sum and difference formulas.

Multiple-Angle Formulas

$$\sin 3\alpha = 3 \sin\alpha - 4 \sin^3\alpha$$

$$\sin 4\alpha = 4 \sin\alpha \cos\alpha - 8 \sin^3\alpha \cos\alpha$$

$$\sin n\alpha = 2 \sin (n-1)\alpha \cos\alpha - \sin (n-2)\alpha$$

$$\cos 3\alpha = 4 \cos^3\alpha - 3 \cos\alpha$$

$$\cos 4\alpha = 8 \cos^4\alpha - 8 \cos^2\alpha + 1$$

$$\cos n\alpha = 2 \cos (n-1)\alpha \cos\alpha - \cos (n-2)\alpha$$

$$\tan 3\alpha = \frac{3 \tan\alpha - \tan^3\alpha}{1 - 3 \tan^2\alpha}$$

$$\tan 4\alpha = \frac{4 \tan\alpha - 4 \tan^3\alpha}{1 - 6 \tan^2\alpha + \tan^4\alpha}$$

$$\tan n\alpha = \frac{\tan(n-1)\alpha + \tan\alpha}{1 - \tan(n-1)\alpha \tan\alpha}$$

Product Formulas

$$\sin\alpha \sin B = \frac{1}{2} \cos(\alpha - B) - \frac{1}{2} \cos(\alpha + B)$$

$$\cos\alpha \cos B = \frac{1}{2} \cos(\alpha - B) + \frac{1}{2} \cos(\alpha + B)$$

$$\cos\alpha \sin B = \frac{1}{2} \sin(\alpha + B) - \frac{1}{2} \sin(\alpha - B)$$

$$\sin\alpha \cos B = \frac{1}{2} \sin(\alpha + B) + \frac{1}{2} \sin(\alpha - B)$$

Sum and Difference Formulas

$$\sin \alpha + \sin B = 2 \sin \left(\frac{\alpha + B}{2}\right) \cos \left(\frac{\alpha - B}{2}\right)$$

$$\sin \alpha - \sin B = 2 \cos \left(\frac{\alpha + B}{2}\right) \sin \left(\frac{\alpha - B}{2}\right)$$

$$\cos \alpha + \cos B = 2 \cos \left(\frac{\alpha + B}{2}\right) \cos \left(\frac{\alpha - B}{2}\right)$$

$$\cos \alpha - \cos B = -2 \sin \left(\frac{\alpha + B}{2}\right) \sin \left(\frac{\alpha - B}{2}\right)$$

LIST 199

Reduction Formulas

Using the formulas below, a circular function of any arc may be expressed in terms of a circular function of an arc between 0 and $\frac{\pi}{4}$.

$$\sin\left(\frac{\pi}{2} - \alpha\right) = \cos \alpha$$

$$\cos\left(\frac{\pi}{2} - \alpha\right) = \sin \alpha$$

$$\tan\left(\frac{\pi}{2} - \alpha\right) = \cot \alpha$$

$$\sin\left(\frac{\pi}{2} + \alpha\right) = \cos \alpha$$

$$\cos\left(\frac{\pi}{2} + \alpha\right) = -\sin \alpha$$

$$\tan\left(\frac{\pi}{2} + \alpha\right) = -\cot \alpha$$

$$\sin\left(\pi - \alpha\right) = \sin \alpha$$

$$\cos\left(\pi - \alpha\right) = -\cos \alpha$$

$$\tan\left(\pi - \alpha\right) = -\tan \alpha$$

$$\sin\left(\pi + \alpha\right) = -\sin \alpha$$

$$\cos\left(\pi + \alpha\right) = -\cos \alpha$$

$$\tan\left(\pi + \alpha\right) = \tan \alpha$$

$$\sin\left(\frac{3\pi}{2} - \alpha\right) = -\cos \alpha$$

$$\cos\left(\frac{3\pi}{2} - \alpha\right) = -\sin \alpha$$

$$\sin\left(\frac{3\pi}{2} + \alpha\right) = -\cos \alpha$$

$$\cos\left(\frac{3\pi}{2} + \alpha\right) = \sin \alpha$$

$$\sin\left(2\pi - \alpha\right) = -\sin \alpha$$

$$\cos\left(2\pi - \alpha\right) = \cos \alpha$$

LIST 200
General Guidelines for Proving Identities

There are several ways of proving proposed trigonometric identities. Usually you will need more than one method for verification. Use the following guidelines.

Before You Begin

1. Know the basic trigonometric identities, and recognize the alternate forms of each.

2. Know procedures for adding and subtracting fractions, reducing fractions, and writing equivalent fractions.

3. Know factoring and special product techniques.

The General Plan

1. Begin with the side that appears more complicated, and try to transform it into the form on the other side.

2. Use substitution and simplification procedures that allow you to work on one side of the equation.

3. Transform each side of the equation independently into the same form.

Hints

1. Avoid substitutions that involve radicals.

2. Use substitutions to change all trigonometric functions into expressions involving only sine and cosine. Then simplify.

3. Clear any fractions in the proposed identity. This may involve multiplying the numerator and denominator of a fraction by the conjugate of each other.

4. Simplify the square root of a fraction by using conjugates to transform it into the quotient of perfect squares.

The Trigonometric Functions in Terms of One Another

An effective method for proving trigonometric identities is to express the trigonometric functions in terms of only sine and cosine. The list below expresses each trigonometric function not only in terms of the sine and cosine, but each other as well.

$$\sin \alpha = \pm \sqrt{1 - \cos^2 \alpha}$$

$$\tan \alpha = \pm \frac{\sin \alpha}{\sqrt{1 - \sin^2 \alpha}}$$

$$\sec \alpha = \pm \frac{1}{\sqrt{1 - \sin^2 \alpha}}$$

$$= \pm \frac{\tan \alpha}{\sqrt{1 + \tan^2 \alpha}}$$

$$= \pm \frac{\sqrt{1 - \cos^2 \alpha}}{\cos \alpha}$$

$$= \frac{1}{\cos \alpha}$$

$$= \pm \frac{1}{\sqrt{1 + \cot^2 \alpha}}$$

$$= \frac{1}{\cot \alpha}$$

$$= \pm \sqrt{1 + \tan^2 \alpha}$$

$$= \pm \frac{\sqrt{\sec^2 \alpha - 1}}{\sec \alpha}$$

$$= \pm \sqrt{\sec^2 \alpha - 1}$$

$$= \pm \frac{\sqrt{1 + \cot^2 \alpha}}{\cot \alpha}$$

$$= \frac{1}{\cos \alpha}$$

$$= \pm \frac{1}{\sqrt{\csc^2 \alpha - 1}}$$

$$= \pm \frac{\csc \alpha}{\sqrt{\csc^2 \alpha - 1}}$$

$$\csc \alpha = \pm \sqrt{1 - \sin^2 \alpha}$$

$$\cot \alpha = \pm \frac{\sqrt{1 - \sin^2 \alpha}}{\sin \alpha}$$

$$\csc \alpha = \frac{1}{\sin \alpha}$$

$$= \pm \frac{1}{\sqrt{1 + \tan^2 \alpha}}$$

$$= \pm \frac{\cos \alpha}{\sqrt{1 - \cos^2 \alpha}}$$

$$= \pm \frac{1}{\sqrt{1 - \cos^2 \alpha}}$$

$$= \pm \frac{\cot \alpha}{\sqrt{1 + \cot^2 \alpha}}$$

$$= \frac{1}{\tan \alpha}$$

$$= \pm \frac{\sqrt{1 + \tan^2 \alpha}}{\tan \alpha}$$

$$= \frac{1}{\sec \alpha}$$

$$= \pm \frac{1}{\sqrt{\sec^2 \alpha - 1}}$$

$$= \pm \sqrt{1 + \cot^2 \alpha}$$

$$= \pm \frac{\sqrt{\csc^2 \alpha - 1}}{\csc \alpha}$$

$$= \pm \sqrt{\csc^2 \alpha - 1}$$

$$= \pm \frac{\sec \alpha}{\sqrt{\sec^2 \alpha - 1}}$$

The sign depends on the quadrant of the angle.

LIST 202

Trigonometric Identities and Oblique Triangles

The following identities are true for any triangle but are most often used to find missing sides and angles of oblique triangles. An *oblique triangle* is a triangle that does not contain a right angle. It may have three acute angles or two acute angles and one obtuse angle. The lengths of the sides are denoted by small letters, and corresponding angles are denoted by capital letters.

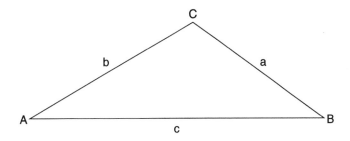

▶ Law of Sines: $\dfrac{a}{\sin A} = \dfrac{b}{\sin B} = \dfrac{c}{\sin C}$

This identity is used if you are given:

- two angles and the side opposite one of them.
- two angles and the included side.
- two sides and the angle opposite one of them. (In this case, there may be no solution, one solution, or two solutions.)

▶ Law of Cosines: $a^2 = b^2 + c^2 - 2bc \cos A$

$$b^2 = c^2 + a^2 - 2ca \cos B$$

$$c^2 = a^2 + b^2 - 2ab \cos C$$

This identity is used if you are given:

- two sides and the included angle.
- three sides.

▶ Law of Tangents:

$$\frac{b-c}{b+c} = \frac{\tan \frac{1}{2}(B-C)}{\tan \frac{1}{2}(B+C)} \qquad \frac{a-b}{a+b} = \frac{\tan \frac{1}{2}(A-B)}{\tan \frac{1}{2}(A+B)} \qquad \frac{c-a}{c+a} = \frac{\tan \frac{1}{2}(C-A)}{\tan \frac{1}{2}(C+A)}$$

This identity may be used if you are given two sides and the included angle, especially if the computations involve logarithms.

LIST 202

(Continued)

▶ Newton's formulas:

$$\frac{b+c}{a} = \frac{\cos\frac{1}{2}(B-C)}{\sin\frac{1}{2}A}$$

$$\frac{c+a}{b} = \frac{\cos\frac{1}{2}(C-A)}{\sin\frac{1}{2}B}$$

$$\frac{a+b}{c} = \frac{\cos\frac{1}{2}(A-B)}{\sin\frac{1}{2}C}$$

▶ Mollweide's formulas:

$$\frac{b-c}{a} = \frac{\sin\frac{1}{2}(B-C)}{\cos\frac{1}{2}A}$$

$$\frac{c-a}{b} = \frac{\sin\frac{1}{2}(C-A)}{\cos\frac{1}{2}B}$$

$$\frac{a-b}{c} = \frac{\sin\frac{1}{2}(A-B)}{\cos\frac{1}{2}C}$$

Both Newton's formulas and Mollweide's formulas are used to check solutions.

LIST 203

Inverse Trigonometric Functions

Since each of the trigonometric functions is periodic, none has an inverse. However, if the domain of each of these functions is restricted to a suitable interval, then each of the functions has an inverse.

These inverse functions are called *inverse trigonometric functions.* Each may be denoted in two ways: by "arc" or $^{-1}$. For example, the inverse of the sine function is the arcsin function or $\sin^{-1}x$. The same notation is used for each of the other five trigonometric functions.

A list of the inverse trigonometric functions, their definitions, domains, and ranges is shown below. Note that *iff* means "if and only if."

Inverse Function	Domain of Inverse	Principal Range of Inverse
$y = \sin^{-1} x$ or arcsin x iff $\sin y = x$	$-1 \leq x \leq 1$	$-\dfrac{\pi}{2} \leq y \leq \dfrac{\pi}{2}$
$y = \cos^{-1} x$ or arccos x iff $\cos y = x$	$-1 \leq x \leq 1$	$0 \leq y \leq \pi$
$y = \tan^{-1} x$ or arctan x iff $\tan y = x$	x is any real number	$-\dfrac{\pi}{2} < y < \dfrac{\pi}{2}$
$y = \cot^{-1} x$ or arccot x iff $\cot y = x$	x is any real number	$0 < y < \pi$
$y = \sec^{-1} x$ or arcsec x iff $\sec y = x$	$|x| \geq 1$	$-\pi \leq y \leq 0,\ y \neq -\dfrac{\pi}{2}$
$y = \csc^{-1} x$ or arccsc x iff $\csc y = x$	$|x| \geq 1$	$-\dfrac{\pi}{2} \leq y \leq \dfrac{\pi}{2},\ y \neq 0$

LIST 204

Vertical and Horizontal Line Tests

Two lines may be drawn on the graph of a relation to determine if the relation is a function and if the function is one-to-one. These tests are called the *vertical line test* and the *horizontal line test.*

Vertical Line Test

▶ If a vertical line intersects the graph no more than once, then the graph represents a function.

Horizontal Line Test

▶ This test determines if a function is one-to-one. A function is one-to-one if for any two elements of the domain of f, $f(x_1) = f(x_2)$ if and only if $x_1 = x_2$.

▶ If a horizontal line intersects the graph of a function no more than once, then the graph represents a one-to-one function.

LIST 205

Relations Between Inverse Trigonometric Functions

The inverse trigonometric functions—$\sin^{-1} x$, $\cos^{-1} x$, $\tan^{-1} x$, $\cot^{-1} x$, $\sec^{-1} x$, and $\csc^{-1} x$—are the inverses of $\sin x$, $\cos x$, $\tan x$, $\cot x$, $\sec x$, and $\csc x$, respectively. The relationships between the inverse functions to each other are shown below, assuming principal values are used.

$$\sin^{-1} x + \cos^{-1} x = \frac{\pi}{2}$$

$$\tan^{-1} x + \cot^{-1} x = \frac{\pi}{2}$$

$$\sec^{-1} x + \csc^{-1} x = \frac{\pi}{2}$$

$$\csc^{-1} x = \sin^{-1} \left(\frac{1}{x} \right)$$

$$\sec^{-1} x = \cos^{-1} \left(\frac{1}{x} \right)$$

$$\cot^{-1} x = \tan^{-1} \left(\frac{1}{x} \right)$$

$$\sin^{-1} (-x) = -\sin^{-1} x$$

$$\cos^{-1} (-x) = \pi - \cos^{-1} x$$

$$\tan^{-1} (-x) = -\tan^{-1} x$$

$$\cot^{-1} (-x) = \pi - \cot^{-1} x$$

$$\sec^{-1} (-x) = \pi - \sec^{-1} x$$

$$\csc^{-1} (-x) = -\csc^{-1} x$$

LIST 206

Graphs of the Inverse Trigonometric Functions

The graphs of the inverse trigonometric relations are those of the corresponding trigonometric functions, except that the roles of x and y are interchanged.

For example, the graph of $y = \arcsin x$ is the graph of $x = \sin y$ and differs from the graph of $y = \sin x$ in that the roles of x and y are switched. The graph of $y = \arcsin x$ is a sine curve drawn on the y-axis.

The graphs below show the graphs of the inverse relations. When the domain is restricted to a suitable interval, as in List 203, "Inverse Trigonometric Functions," the graph is a function. The graphs of the principal values of the inverse functions are shown in bold.

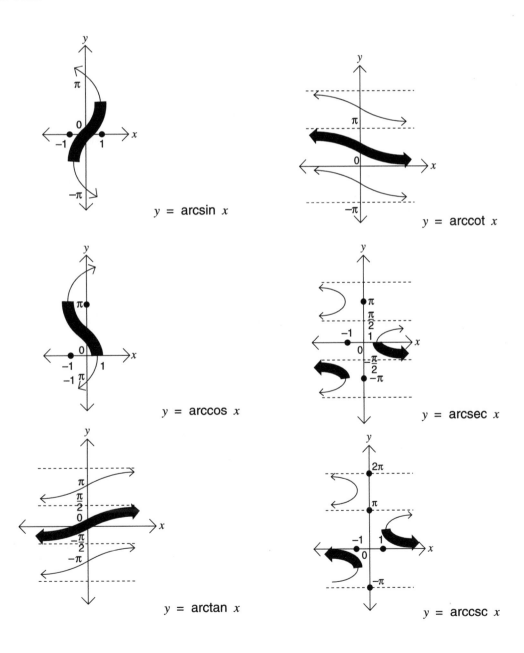

LIST 207

Limits

The notion of a limit is fundamental to the study of calculus. In everyday life, we use expressions such as "the speed limit," "my patience is limited," and "the only limits to your potential are those you impose yourself." These expressions suggest that a limit is a type of boundary that may be reached on some occasions, but on some others may be exceeded.

In calculus, a limit is defined: if $f(x)$ becomes arbitrarily close to a single number L as x approaches c from either side, then we write:

$$\lim_{x \to c} f(x) = L$$

and say that the limit of $f(x)$ as x approaches c is L.

Note that "x approaches c" means that no matter how close x comes to the value of c, there is another value of x (different from c) in the domain of f that is even closer to c.

One-Side Limits

The definition of the limit $\lim_{x \to c} f(x) = L$ requires that x approaches c from the right as well as the left.

▶ $x \to c-$ is read "x approaches c from the left."

▶ $x \to c^+$ is read "x approaches c from the right."

Each limit obtained from the one-sided approach has a special name:

▶ $\lim_{x \to c-} f(x)$ is called the *limit from the left.*

▶ $\lim_{x \to c+} f(x)$ is called the *limit from the right.*

Properties of Limits

The properties hold for real numbers b and c. n is a positive integer.

▶ $\lim_{x \to c} b = b$

▶ $\lim_{x \to c} x = c$

▶ $\lim_{x \to c} x^n = c^n$

▶ $\lim_{x \to c} \sqrt[n]{x} = \sqrt[n]{c}, \quad c > 0$

LIST 208

Operations with Limits

In order for the limit of a function to exist as $x \to c$, it must be true that both one-sided limits exist and are equal. If f is a function and c and L are real numbers, then

$$\lim_{x \to c} f(x) = L$$

if and only if

$$\lim_{x \to c-} f(x) = L \text{ and } \lim_{x \to c+} f(x) = L.$$

If f and g are two functions, we can produce several new functions: their sum, $f + g$; difference, $f - g$; product, fg; and quotient, $\dfrac{f}{g}$. The limits of these functions as $x \to c$ can easily be determined once we know the limit of f at c and the limit of g at c.

Rules governing the limits of functions follow. b and c are real numbers. n is a positive integer. If the limits of $f(x)$ and $g(x)$ exist as x approaches c, then:

Constant Multiple: $\lim\limits_{x \to c} \left[bf(x) \right] = b \left[\lim\limits_{x \to c} f(x) \right]$

Addition: $\lim\limits_{x \to c} \left[f(x) + g(x) \right] = \lim\limits_{x \to c} f(x) + \lim\limits_{x \to c} g(x)$

Subtraction: $\lim\limits_{x \to c} \left[f(x) - g(x) \right] = \lim\limits_{x \to c} f(x) - \lim\limits_{x \to c} g(x)$

Multiplication: $\lim\limits_{x \to c} \left[f(x)\, g(x) \right] = \left[\lim\limits_{x \to c} f(x) \right] \left[\lim\limits_{x \to c} g(x) \right]$

Division: $\lim\limits_{x \to c} \dfrac{f(x)}{g(x)} = \dfrac{\lim\limits_{x \to c} f(x)}{\lim\limits_{x \to c} g(x)}$ if $\lim\limits_{x \to c} g(x) \neq 0$

Power: $\lim\limits_{x \to c} \left[f(x) \right]^{n} = \left[\lim\limits_{x \to c} f(x) \right]^{n}$

Radical: $\lim\limits_{x \to c} \sqrt[n]{f(x)} = \sqrt[n]{\lim\limits_{x \to c} f(x)},\quad \lim\limits_{x \to c} f(x) > 0$

LIST **209**

Continuity

A *function is continuous* at $x = c$ if there is no interruption of the graph of f at c. There are no holes, jumps, or gaps. The graph of a function is continuous (on an interval) if its graph (on the interval) can be traced without lifting the pencil from the paper.

Note the following definitions:

▶ *Continuity at a Point:* A function is said to be continuous at c if the following conditions are met:

- ▪ $f(c)$ is defined.
- ▪ $\lim\limits_{x \to c} f(x)$ exists.
- ▪ $\lim\limits_{x \to c} f(x) = f(c)$.

▶ *Continuity on an Open Interval:* A function is said to be continuous on an interval (a,b) if it is continuous at each point in the interval.

▶ *Continuity on a Closed Interval:* If f is continuous on the open interval (a,b) and $\lim\limits_{x \to a_+} f(x) = f(a)$ and $\lim\limits_{x \to b^-} f(x) = f(b)$, then f is continuous on $[a,b]$, and we say f is continuous from the right at a and continuous from the left at b, provided f is defined on $[a,b]$.

LIST **210**

Properties of Continuous Functions

Because continuity is defined in terms of limits, continuous functions and limits share many of the same properties. In the list below, f and g are continuous at c, and k is a real number. The following functions are also continuous at c.

▶ The sum is $f + g$.
▶ The difference is $f - g$.
▶ The constant multiple is kf.
▶ The product is fg.
▶ The quotient is $\dfrac{f}{g}$, $g(c) \neq 0$.
▶ The composition is $f \circ g(x)$ where g is continuous at c and f is continuous at $g(c)$.
▶ The polynomial function is continuous at every number.
▶ The rational function is continuous at every number in its domain.
▶ $f(x) = \sqrt[n]{x}$ is continuous at every number if n is an odd positive integer.
▶ $f(x) = \sqrt[n]{x}$ is continuous at every positive number if n is an even positive integer.

LIST 211

Differential Rules

Differentiation is the process for finding the rate at which one quantity changes with respect to another. The rules needed to differentiate algebraic functions are listed below. u and v are differentiable functions of x, and y is a differentiable function of u. c is a constant. n is a real number.

Constant Rule: $$\frac{d}{dx}[c] = 0$$

Constant Multiple Rule: $$\frac{d}{dx}[cu] = c\frac{du}{dx}$$

Sum Rule: $$\frac{d}{dx}[u + v] = \frac{du}{dx} + \frac{dv}{dx}$$

Difference Rule: $$\frac{d}{dx}[u - v] = \frac{du}{dx} - \frac{dv}{dx}$$

Product Rule: $$\frac{d}{dx}[uv] = u\frac{dv}{dx} + v\frac{du}{dx}$$

Quotient Rule: $$\frac{d}{dx}\left[\frac{u}{v}\right] = \frac{v\frac{du}{dx} - u\frac{du}{dx}}{v^2}$$

Power Rules: $$\frac{d}{dx}\left[x^n\right] = nx^{n-1}$$

$$\frac{d}{dx}\left[u^n\right] = nu^{n-1}\frac{du}{dx}$$

Chain Rule: $$\frac{dy}{dx} = \frac{dy}{du} \cdot \frac{du}{dx}$$

The following rules are special cases of the Power Rules or combinations of the Constant Multiple and Power Rules:

$$\frac{d}{dx}[x] = 1$$

$$\frac{d}{dx}\left[cx^n\right] = cnx^{-1}$$

$$\frac{d}{dx}[cx] = c$$

Higher-Order Derivatives

Since the derivative of a function is a function, you may find the derivative of the derivative (called the second derivative). By repeating this process, you can obtain higher-order derivatives.

To find the higher-order derivatives of $f(x)$, do the following:

1. Find the derivative of a function. This is called finding the first derivative. It is denoted $f'(x)$ and is read "f prime of x."

2. Find the derivative of the first derivative. This is called finding the second derivative and is denoted $f''(x)$. It is read "f double prime of x."

3. Find the derivative of the second derivative, which is called finding the third derivative. It is denoted $f'''(x)$ and is read "f triple prime of x."

4. Continue this process, provided each successive derivative is differentiable.

Commonly used notation for higher-order derivatives is shown below.

First Derivative	*Second Derivative*	*Third Derivative*	*Fourth Derivative*	*nth Derivative*
$f'(x)$	$f''(x)$	$f'''(x)$	$f^{(4)}(x)$	$f^{(n)}(x)$
y'	y''	y'''	$y^{(4)}$	$y^{(n)}$
$\dfrac{dy}{dx}$	$\dfrac{d^2 y}{dx^2}$	$\dfrac{d^3 y}{dx^3}$	$\dfrac{d^4 y}{dx^4}$	$\dfrac{d^n y}{dx^n}$
$\dfrac{d}{dx}\big[f(x)\big]$	$\dfrac{d^2}{dx^2}\big[f(x)\big]$	$\dfrac{d^3}{dx^3}\big[f(x)\big]$	$\dfrac{d^4}{dx^4}\big[f(x)\big]$	$\dfrac{d^n}{dx^n}\big[f(x)\big]$
$Dx(y)$	$Dx^2(y)$	$Dx^3(y)$	$Dx^4(y)$	$Dx^n(y)$

LIST 213

Curve Drawing: Application of Higher-Order Derivatives

A function $y = f(x)$ is increasing if its graph "moves up" as x moves to the right. It is decreasing if its graph "moves down" as x moves to the right. This can be determined by finding the first derivative of the function, provided the function is differentiable.

▶ If the first derivative is positive, the value of the original function is increasing.

▶ If the first derivative is negative, the value of the original function is decreasing.

▶ If the first derivative is 0, the original curve has a horizontal tangent at that point.

The second derivative (provided it exists) is used to determine concavity and point of inflection. A *point of inflection* is the point where the graph of a continuous function possesses a tangent line and its concavity changes from upward to downward or downward to upward.

▶ If the second derivative is positive, then the original curve is concave upward (shaped like a ∪).

▶ If the second derivative is negative, then the original curve is concave downward (shaped like a ∩).

▶ If the second derivative is 0, then the original curve has a point of inflection, provided that the second derivative is positive on one side of the point and negative on the other.

The first and second derivatives (if they exist) may be used to test for relative minimum and relative maximum. *Relative minimum* and *relative maximum* are the lowest and highest points on a given interval.

▶ If the first derivative is 0 and
  ▪ the second derivative is positive, the point is a relative minimum.
  ▪ the second derivative is negative, the point is a relative maximum.
  ▪ the second derivative is 0, the test fails.

LIST 214

Notation for Antiderivatives

Antidifferentiation is the process of determining the original function from its derivative. Note the following:

▶ A function F is called an antiderivative of a function f if for every x in the domain of f, $F'(x) = f(x)$.

▶ $F(x)$ is an antiderivative of $f(x)$. It is used synonymously with F and is an antiderivative of f.

▶ The antidifferentiation process is also called *integration*.

▶ $\int$ is the integral sign.

▶ $\int f(x)\,dx$ is called the indefinite integral of $f(x)$ and denotes a family of antiderivatives of $F(x)$.

▶ $\int f(x)\,dx = F(x) + C$ means that F is an antiderivative of f. That is $F'(x) = f(x)$ for all x in the domain of f. $f(x)$ is called the *integrand*. C is the *constant of integration*.

LIST 215

Integration Rules

Because integration and differentiation are inverse relations, integration formulas may be obtained directly from differentiation formulas. They are summarized below.

Constant Rule:	$\int k\,dx = kx + c$
Constant Multiple Rule:	$\int kf(x)\,dx = k \int f(x)\,dx$
Sum Rule:	$\int \left[f(x) + g(x) \right] dx = \int f(x)\,dx + \int g(x)\,dx$
Difference Rule:	$\int \left[f(x) - g(x) \right] dx = \int f(x)\,dx - \int g(x)\,dx$
Power Rule:	$\int x^n\,dx = \dfrac{x^{n+1}}{n+1} + c, \quad n \neq -1$
General Power Rule:	$\int u^n \dfrac{du}{dx}\,dx = \dfrac{u^{n+1}}{n+1} + c, \quad n \neq -1$

Note that u is a differentiable function of x.

LIST 216

Definite Integrals

Definite integrals are integrals that have upper and lower limits. Note the following facts.

▶ $\int_a^b f(x)\,dx$ is called the definite integral from a to b where

  a is the lower limit of integration.

  b is the upper limit of integration.

▶ The Fundamental Theorem of Calculus describes a means for evaluating a definite integral, which is stated as follows:

  If a function f is continuous on the interval $[a,b]$, then

  $\int_a^b f(x)\,dx = F(b) - F(a)$ where F is any function such that $F'(x) = f(x)$ for all x in $[a,b]$.

▶ Properties of definite integrals are listed below. f and g are integrable on $[a,b]$.

  $\int_a^b kf(x)\,dx = k\int_a^b f(x)\,dx$; k is a constant.

  $\int_a^b \left[f(x) + g(x)\right]dx = \int_a^b f(x)\,dx + \int_a^b g(x)\,dx$

  $\int_a^b \left[f(x) - g(x)\right]dx = \int_a^b f(x)\,dx - \int_a^b g(x)\,dx$

  $\int_a^b f(x)\,dx = \int_a^c f(x)\,dx + \int_c^b f(x)\,dx$ if $a < c < b$

▶ $\int_a^b f(x)\,dx = \int_a^c f(x)\,dx + \int_c^b f(x)\,dx$ regardless of the order of a, b, and c.

▶ If k is any constant, then $\int_a^b k\,dx = k(b-a)$

  $\int_a^a f(x)\,dx = 0$

  $\int_a^b f(x)\,dx = -\int_b^a f(x)\,dx$

▶ The definite integral $\int_a^b f(x)\,dx$ is a number, whereas the indefinite integral $\int f(x)\,dx$ is a family of functions.

LIST **217**

Derivatives and Integrals of Trigonometric Functions

The list below shows the derivatives and corresponding integrals of trigonometric functions. (See List 183, "Right Triangle Definitions of Trigonometric Functions," and List 186, "Circular Function Definition of Trigonometric Functions.")

$$\frac{d}{dx}[\sin u] = \cos u \frac{du}{dx} \qquad\qquad \int \cos u \; du = \sin u + C$$

$$\frac{d}{dx}[\cos u] = -\sin u \frac{du}{dx} \qquad\qquad \int \sin u \; du = -\cos u + C$$

$$\frac{d}{dx}[\tan u] = \sec^2 u \frac{du}{dx} \qquad\qquad \int \sec^2 u \; du = \tan u + C$$

$$\frac{d}{dx}[\cot u] = -\csc^2 u \frac{d}{dx} \qquad\qquad \int \csc^2 u \; du = -\cot u + C$$

$$\frac{d}{dx}[\sec u] = \sec u \tan u \frac{du}{dx} \qquad\qquad \int \sec u \tan u \; du = \sec u + C$$

$$\frac{d}{dx}[\csc u] = -\csc u \cot u \frac{du}{dx} \qquad\qquad \int \csc u \cot u \; du = -\csc u + C$$

The integrals of the six basic trigonometric functions follow.

$$\int \sin u \; du = -\cos u + C$$

$$\int \cot u \; du = \ln|\sin u| + C$$

$$\int \cos u \; du = \sin u + C$$

$$\int \sec u \; du = \ln|\sec u + \tan u| + C$$

$$\int \tan u \; du = -\ln|\cos u| + C$$

$$\int \csc u \; du = \ln|\csc u - \cot u| + C$$

LIST 218

Exponential Functions

Many functions that have been studied involve a variable raised to a constant power such as $f(x) = x^2$, $k(x) = \sqrt{x} = x^{\frac{1}{2}}$, or $g(x) = x^{-1}$. If the roles are interchanged and a constant is raised to a variable power, then a group of functions called *exponential functions* results. Some examples are $f(x) = 2^x$, $k(x) = \left(\frac{1}{2}\right)^x$, and $g(x) = 3^{-x}$. Any positive base except 1 can be used for exponential functions. Note the following about exponential functions.

▶ If $a > 0$ and $a \neq 1$, then the exponential function with base a is given by $y = a^x$.

▶ In calculus, the choice for the base is e, an irrational number whose decimal approximation is $e \approx 2.71828 \ldots$

▶ e is defined by a limit definition: $e = \lim_{x \to 0} (1 + x)^{\frac{1}{x}}$.

▶ Properties of exponential functions include the following:

$$e^0 = 1$$
$$e^a \cdot e^b = e^{a+b}$$
$$\frac{e^a}{e^b} = e^{a-b}$$
$$\left(e^a\right)^b = e^{ab}$$

▶ Derivatives of exponential functions include $\frac{d}{dx}\left[e^x\right] = e^x$ and $\frac{d}{dx}\left[e^u\right] = e^u \frac{du}{dx}$ where u is a differentiable function of x.

▶ Integrals of exponential functions include $\int e^x \, dx = e^x + c$ and $\int e^u \frac{du}{dx} \, dx = e^u +$ where u is a differentiable function of x.

Natural Logarithmic Functions

The exponential function $f(x) = e^x$ is both increasing and continuous. It possesses an inverse called the *natural logarithmic function,* which is defined below.

- $\ln x = b$ if and only if $e^b = x$. ($\ln x$ is read "the natural log of x.")

- The inverse properties of $\ln x$ and e^x are $\ln e^x$ and $e^{\ln x} = x$.

- The properties of logarithms include:

  $\ln 1 = 0$

  $\ln xy = \ln x + \ln y$

  $\ln \dfrac{x}{y} = \ln x - \ln y$

  $\ln x^y = y \ln x$

- Derivatives of natural logarithmic functions include

  $\dfrac{d}{dx}[\ln x] = \dfrac{1}{x}$ and $\dfrac{d}{dx}[\ln u] = \dfrac{1}{u}\dfrac{du}{dx}$.

  u is a differentiable function of x.

- The log rule for integration is

  $\displaystyle\int \dfrac{1}{u}\dfrac{du}{dx}\,dx = \int \dfrac{1}{u}\,du = \ln|u| + c$.

  u is a differentiable function of x.

Section Six

Math in Other Areas

Descriptive Statistics

Descriptive statistics is a method of collecting, organizing, analyzing, and using numerical data. Suppose a group of test scores of a class of 10 students was 80, 70, 90, 100, 30, 100, 100, 90, 70, and 60. The group can be described statistically, as noted below.

Data—facts and figures collected on the same characteristics of a population or sample. In this case, one or more of the scores is an example of the data.

Population—all of the members of a group or item. In the example, the population is all 10 scores.

Outlier—one or more values that appears unusually large or small and out of place when compared with the other data of a set. 30 is an outlier in the test scores in the example above.

Frequency—the number of times a score or group of scores occurs. The frequency of 30 is 1; 60 is 1; 70 is 2; 80 is 1; 90 is 2; 100 is 3.

Mean—the average of all scores, denoted by *x*. Adding the scores and dividing by 10 gives a mean of 79. (The mean is also called the *average*.)

Median—the number in the middle when the numbers are ordered from least to greatest or greatest to least. For the test scores of the example, the median is shown as:

30, 60, 70, 70, 80, 90, 90, 100, 100, 100

Since there is no single "middle" number, the average of the two "middle" numbers is taken: $(80 + 90) \div 2 = 85$. Thus, 85 is the median.

Mode—the number that appears most often in a group. In the example, 100 is the mode because it occurs three times, which is the most of any other number.

Range—the highest score minus the lowest score. $100 - 30 = 70$. Therefore, 70 is the range.

Measures of Central Tendency

Mean, median, and *mode* are often referred to as the *measures of central tendency*. While it is hard to determine which is the most important to use when describing specific data, here are some guidelines:

▶ The mode is usually most appropriate for evaluating the quality of a program or situation.

▶ The median is usually used when the sample is small and the data extreme.

▶ The mean is usually used when the sample is large.

▶ Both mode and median are usually unaffected by extreme data.

▶ When the range is small, the mean is usually the best measure of central tendency.

LIST 221

Types of Graphs, Displays, and Tables

We live in a world where we are bombarded by data. One of the best ways to organize and represent information is graphically.

Artistic Graph—displays information pictorially. Unlike a pictograph, which represents data in the form of symbols, an artistic graph uses an illustration to show information.

Bar Graph—uses the lengths of bars to compare data. Bars are separated by spaces. There are single bar graphs, double bar graphs, and even triple bar graphs.

Box-and-Whisker Plot—displays the median of a set of data, the median of each half of the data, and the least and greatest values of the data. Box-and-whisker plots are used to compare sets of data.

Circle Graph—represents data expressed as parts of a whole. The circle equals 100%. Each part of a circle graph is called a *sector*. Circle graphs are also known as *pie graphs.*

Frequency Polygon—represents frequencies. A type of line graph, it is made by connecting the midpoints of the tops of the bars of a histogram.

Frequency Table—organizes a set of data. Frequency tables show the number of times each item of the set occurs.

Histogram—displays the frequency of data that has been organized into equal distributions. It is a special kind of bar graph.

Line Graph—displays data as points that are connected by line segments. Single and double line graphs are common.

Line Plot—displays information on a number line in the form of a vertical graph.

Pictograph—uses a symbol or symbols to represent data. Values are approximations.

Scattergram—displays relationships among data. Information is represented by unconnected points. A given number may have more than one point on either scale. Points seldom lie in a line.

Stem-and-Leaf Plot—represents data where each number is displayed by a stem and a leaf. Stem-and-leaf plots show the value for each piece of datum. They may be single or back to back.

Trend Line—makes predictions. It is a line that can be drawn near the points on a scattergram. It is also called the *line of best fit.*

Topics in Discrete Math

Discrete mathematics is a branch of mathematics that helps students to understand and interpret data. It includes various strategies and methods for solving problems based on real-world situations. The study of discrete math includes but is not limited to the following topics.

- ▶ Multiplication principle of counting

- ▶ Organized lists, charts, tree diagrams, tables, and Venn diagrams

- ▶ Combinations

- ▶ Permutations

- ▶ Pascal's Triangle

- ▶ Factorials

- ▶ Vertex-edge graphs

- ▶ Scheduling problems/conflict avoidance

- ▶ Routes and circuits

- ▶ Graph coloring

- ▶ Apportionment

- ▶ Strategies for making fair decisions

Permutations

Permutations are arrangements of things in a particular order. The symbolic representation of the total number of arrangements of *n* things taken *r* at a time is $_nP_r$.

Permutation Formulas and Applications

▶ A *factorial* is a product of all whole numbers less than or equal to a given number and is represented as *n*!

▶ $n! = n(n-1)(n-2)(n-3) \ldots (3)(2)(1)$. For example, $6! = (6)(5)(4)(3)(2)(1) = 720$.

▶ $0! = 1$

▶ $1! = 1$

▶ To find the total number of arrangements of *n* things taken *r* at a time, use this formula:

$$_nP_r = \frac{n!}{(n-r)!}$$

Application: Suppose you wanted to rearrange the letters of the word "MATH" to form other four-letter arrangements. How many are there? You could list them and find 24 arrangements:

MATH	AMTH	TMAH	HMAT
MAHT	AMHT	TMHA	HMTA
MTAH	ATHM	TAMH	HAMT
MTHA	ATMH	TAHM	HATM
MHAT	AHMT	THAM	HTAM
MHTA	AHTM	THMA	HTMA

Or you could use the formula $_nP_r$ where $n = 4$ and $r = 4$ (four letters taken four at a time).

$$_4P_4 = \frac{4!}{(4-4)!} = \frac{4 \cdot 3 \cdot 2 \cdot 1}{0!} = 24$$

▶ P_c is the total number of all circular permutations of *n* things arranged in a circle.

$$P_c = (n-1)!$$

Application: Suppose you wanted to arrange the letters of MATH in a circle. In a circle there is no first, second, third, or any other position. If one position is fixed as a reference point, then the other three letters may be permutated around it. There are six arrangements, all of which are shown on the next sheet.

LIST 223

(Continued)

M	M	M	M	M	M
A H	T H	H T	A T	T A	H A
T	A	A	H	H	T

Or you could use the formula $P_c = (n-1)!$ where $n = 4$.

$$P_c = (4-1)! = 3! = (3)(2)(1) = 6$$

▶ P_a is the total number of indistinguishable permutations possible for n things when n_1 are alike and n_2 are alike and so on.

$$P_a = \frac{n!}{n_1!n_2!\dots}$$

Application: Suppose you want to find all possible arrangements of the letters in the word THAT. Two letters are indistinguishable: the two Ts. You could make a list:

THAT	HATT	ATHT
THTA	HTAT	ATTH
TAHT	HTTA	AHTT
TATH		
TTAH		
TTHA		

Or you could use the formula $P_a = \frac{n!}{n_1!n_2!\dots}$ where $n = 4$ and $n_1 = 2$.

$$P_a = \frac{4!}{2!} = \frac{4 \cdot 3 \cdot 2 \cdot 1}{2 \cdot 1} = 12$$

LIST 224

Combinations

Combinations are a grouping of items without regard to order. The symbolic representation of the total combinations of n things taken r at a time is $_nC_r$.

Combinations Formulas and Applications

▶ $_nC_r = \dfrac{n!}{r!(n-r)!}$ *Note:* $n! = n(n-1)(n-2)\ldots(3)(2)(1)$

▶ $_nC_r = \dfrac{_nP_r}{r!}$ *Note:* $_nP_r = \dfrac{n!}{(n-r)!}$

▶ $_nC_n = \dfrac{n!}{n!} = 1$

Example: How many ways may groups of 3 students be chosen from a group of 10? The grouping of 3 students is important, and order is not. In this case, $n = 10$ and $r = 3$.

You may use the following formula:

$$_nC_r = \frac{n!}{r!(n-r)!} = \frac{10!}{3!(10-3)!} = \frac{10\cdot9\cdot8}{3\cdot2\cdot1} = \frac{720}{3} = 120$$

Or you may use the following formula, which relates combinations to permutations.

$$_nC_r = \frac{_nP_r}{r!} = \frac{\frac{10!}{(10-3)!}}{3!} = \frac{10\cdot9\cdot8}{3\cdot2\cdot1} = 120$$

Intergroup combination formulas are shown below.

▶ $N = {_{n_1}C_{r_1}} \cdot {_{n_2}C_{r_2}} \cdot \ldots \cdot {_{n_k}C_{r_k}}$ is the total number of joint selections possible from k different sets of $n_1, n_2, \ldots, n_k$ things taken $r_1, r_2, \ldots, r_k$ at a time, respectively.

▶ $N = n_1 n_2 n_3 \ldots n_k$ is a special case where $r_1 = r_2 = r_k = 1$.

▶ $N = n^k$ is a special case where $n_1 = n_2 = n_k = n$.

To apply the intergroup combination formulas, try this question. How many groups of 3 students may be chosen from 4 members of Room 101 and 5 members of Room 102? In this case,

$n_1 = 4$ (possibilities from Room 101)

$n_2 = 5$ (possibilities from Room 102)

$r_1 = r_2 = 3$ (because you are forming groups of 3 members)

$$N = {_{n_1}C_{r_1}} \cdot {_{n_2}C_{r_2}} = {_4C_3} \cdot {_5C_3} = \frac{4!}{3!(4-3)!} \cdot \frac{5!}{3!(5-3)!} = \frac{4}{1} \cdot \frac{5\cdot4}{2\cdot1} = 40$$

▶ $M = {_1C_k} + {_2C_k} + \ldots {_kC_k} = 2^k - 1$ where M is the sum of all possible combinations of k things taken 1, 2, 3, ... and any other number up to and including k, at a time. For example, how many different selections can be made of 1 or more books from a shelf of 5 books? $k = 5$ (because you are choosing 5 books).

$$M = 2^k - 1 = 2^5 - 1 = 32 - 1 = 31$$

Probability and Odds

Probability theory is the branch of mathematics that deals with the chances of specific events happening. By assigning measures that can be computed and compared, it tries to take the randomness out of prediction.

Definitions

▶ *Experiment*—a procedure that has the same possible outcomes every time it is repeated. No single outcome is predictable.

Example: If you pick a card randomly from a deck of cards, you know you will choose a card, but it is impossible to predict what card you will pick.

▶ *Sample Space*—the set of all possible outcomes.

Example: In the deck of cards, the sample space for picking 1 card is 52, because a complete deck of cards consists of 52 cards.

▶ *Event*—a subset of a sample space.

Example: In choosing a card from a deck of 52 cards, an event is picking the two of hearts.

▶ *Independent Events*—two separate events such that the outcome of one does not affect the outcome of the other.

Example: If you choose the ace of spades, return it to the deck, and then choose the six of hearts, the events are independent.

▶ *Dependent Events*—two separate events such that the outcome of one does affect the outcome of the other.

Example: If you choose the ace of spades, return it, and choose a card of the same suit, the events are dependent. (The suit of the second card depends on the suit of the first.)

▶ *Mutually Exclusive Events*—two events that cannot both occur at the same time.

Example: Choosing an ace of spades and a two of hearts when you are only picking *one* card.

LIST 225
(Continued)

Probability Laws

▶ Probability of *A* is denoted as *P(A)*, which equals

$$\frac{\text{the number of outcomes of Event } A}{\text{the number of possible outcomes}}$$

as long as all outcomes are equally likely. *Example:*

$$P(\textit{ace of spades}) = \frac{1}{52} = \frac{1 \text{ card}}{\text{total outcomes}}$$

▶ Probability of *A* not occurring is denoted as

$$P(\textit{not A}) = 1 - P(A)$$

as long as all outcomes are equally likely. *Example:* The probability of not picking an ace of spades is $1 - \frac{1}{52} = \frac{51}{52}$.

▶ $0 \leq$ probability of $A \leq 1$. If an event will definitely happen, then the probability is 1. If an event is impossible, then the probability is 0.

▶ The sum of all the probabilities in any experiment must equal 1.

▶ Probability of *A* and *B* is denoted as *P(A and B) = P(A, given B)P(B) = P(A)P(B)* if *A* and *B* are independent events. *Example:* The probability of choosing an ace of spades, returning it, and then picking a two of hearts is $\frac{1}{52} \cdot \frac{1}{52} = \frac{1}{2704}$ because the events are independent.

▶ Probability of *A* or *B* is denoted as *P(A or B) = P(A) + P(B) − P(A and B) = P(A) + P(B)* if *A* and *B* are mutually exclusive events. *Example:* The probability of picking an ace of spades or a two of hearts is $\frac{1}{52} + \frac{1}{52} = \frac{2}{52} = \frac{1}{26}$.

Odds of Event *A* are denoted as

$$\frac{P(A)}{P(\textit{not A})}$$

Example: The odds of choosing an ace of spades is

$$\frac{P(\textit{choosing the ace of spades})}{P(\textit{not choosing the ace of spades})} = \frac{\frac{1}{52}}{\frac{51}{52}} = \frac{1}{51}$$

LIST **226**

Odds in Poker

We are all familiar with the shrewd poker player who bluffs his opponents and wins the big hand. Without question, bluffing is a major part of the game, but poker is, above all else, a game of chance.

Every time a game of poker is played, a player may receive any one of a possible 2,598,960 hands. The odds of being dealt a particular hand are a result of mathematics. Knowing the odds helps a player decide when to up the pot or fold.

The following hands are listed in descending order. A full house, for example, beats a flush but does not beat four of a kind.

Hand	*Definition*	*Odds Against*
Royal Flush	sequence of ten to ace in the same suit	649,739 to 1
Straight Flush	sequence of five cards in the same suit	72,192 to 1
Four of a Kind	four of the same card; four 8s, for example	4,164 to 1
Full House	three of a kind and one pair	693 to 1
Flush	five cards of the same suit but not in sequence	508 to 1
Straight	five cards in sequence but of mixed suit	254 to 1
Three of a Kind	three of the same card; for example, three 10s	46 to 1
Two Pairs	two sets of the same cards; for example, two queens and two 5s	20 to 1
One Pair	two of the same card; two 9s, for example	1.37 to 1
Empty Hand	no winners	1 to 1

LIST 227

Odds in Dice

People have been playing dice for thousands of years. Dice are small cubes that have one, two, three, four, five, or six dots on their faces. The dots indicate the numbers 1 through 6.

One of the best-known dice games is craps. The thrower makes a money bet, which is matched, or covered, by one or more opponents. The thrower then tosses two dice. If the thrower rolls a 7 or 11, he or she wins. If the thrower tosses 2, 3, or 12, he or she loses. If any other number results, that number becomes the thrower's point, and he or she continues to throw until the same number is thrown, which then wins. If the thrower instead tosses a 7, he or she loses the bet as well as the toss.

Following are the odds on dice for one throw.

Dice Total	Odds Against
2	35 to 1
3	17 to 1
4	11 to 1
5	8 to 1
6	31 to 5
7	5 to 1
8	31 to 5
9	8 to 1
10	11 to 1
11	17 to 1
12	35 to 1

Copyright © 2005 by Judith A. Muschla and Gary Robert Muschla

LIST **228**

The Chances Are . . .

Have you ever wondered what the odds are that you might be struck by lightning? Or that you might win a lottery? Or you might be involved in a car accident? Or one day become a movie star? Following are your chances for a variety of occurrences.

Event	*Odds*
Being struck by lightning	1 in 600,000
Starring in a movie	1 in 385,000
Winning a major lottery	1 in 5,200,000
Dying in a plane crash	1 in 10,000,000
Dying in a train accident	Less than 1 in 1,000,000
Being hurt in a car accident	1 in 75
Wearing glasses or contacts at some time in your life	1 in 2
Someday believing you saw a UFO	1 in 10
Having your marriage end in divorce	1 in 2
Someday having a too high cholesterol level	1 in 4
Someday having high blood pressure	2 in 5
Having a supernatural experience	1 in 15
The chances of your new business still being in business after five years	1 in 2
If you are a golfer, sinking a hole-in-one	1 in 10,700
If you are a baseball pitcher, pitching a no-hitter	1 in 1,300
If you are pregnant woman, your chances of giving birth to twins	1 in 50

LIST 229

Ways People Are Paid

People are paid for the work they do in various ways (although, of course, never enough). The following list provides the ways people are compensated for their work. Bartering, developed during prehistoric times, was probably one of the first methods of exchange, but some of the other methods have been around almost as long.

► *Salary*—a fixed rate of payment for services on a regular basis. A salary may be paid weekly, biweekly (two times per month), or monthly.

► *Commission*—payment based on a percentage of sales.

► *Salary plus commission*—payment that is a combination of a set amount (salary) and a percentage of sales (commission).

► *Hourly wage*—a fixed payment based on the amount of hours worked.

► *Overtime*—payment for work done in addition to one's regular hours.

► *Fee*—a fixed payment based on a particular job; for example, a fee earned for painting a house.

► *Bonus*—an extra payment for a job well done or a job effort that surpasses what is considered adequate. Sometimes bonuses are given before a major holiday.

► *Merit pay*—payment for reaching a specific goal; similar to a bonus.

► *Piecework*—payment for the number of products completed.

► *Differential piecework*—a payment plan similar to piecework, except that workers are paid more as they move to higher levels of production.

► *Tip*—a small amount of money given in recognition of service rendered; also called a *gratuity*. For many services, a standard tip amounts to 15% of the bill for the service.

► *Royalties*—a share paid to an author or composer out of the proceeds resulting from the sale or performance of the work. A royalty may also be the share paid to an inventor or proprietor for the use of his or her invention or services. Grantors of mineral leases may be paid royalties for the use of their land.

► *Bartering*—an exchange of services or products agreed on by two people. No money exchanges hands.

Deductions

When you are paid by check for the work you do, the amount you receive is often less than the amount you earned. The amount you earned is your *gross pay*. The amount you actually receive is called your *net pay* (or *take-home pay*). The difference is a result of payroll deductions, of which the most common are listed below.

▶ *Federal Income Withholding Tax*—a method by which the U.S. government collects a portion of the income tax it anticipates you will owe based on your salary and the number of dependents you claim on your W-4 withholding form.

▶ *Social Security Tax (F.I.C.A.)*—the tax levied to sustain the Social Security system.

▶ *State Income Tax*—for those who live in a state that levies an income tax.

▶ *City Income Tax*—for those who live in a city that levies an income tax.

▶ *Other Possible Deductions*
- Employee's contribution to health, life, and/or disability insurance premiums
- Union or agency dues
- Pension plans
- Credit union dues
- Savings plans or stock plans
- Pension loans

▶ *Less Typical Deductions*
- Lost work time
- Court liens

To determine that your net pay is correct, do the following:

1. Write down the amount of each deduction.
2. Make sure that the deduction applies to your check.
3. Add to find the total amount of deductions.
4. Subtract your answer from your gross pay for the given pay period. Your answer here should equal your net pay.

LIST 231

How to Write and Endorse a Check

Checks are a common method of monetary payment, generally accepted in place of cash. The steps for writing and endorsing a check are shown here and on the next page.

How to Write a Check

▶ Follow these general guidelines:

- Use a nonerasable pen.

- Write clearly.

- Do not leave any blank spaces.

- Do not cross out or change any part of the check unless you mark the change with your initials.

- Record the check in the check register or on the check stub.

- Make sure that you have enough money in the account, or the check will not be honored by the bank. (In other words, the check will bounce.)

▶ Fill out the information on the front (or face) of the check:

- *Date.* Write the date the check is written.

- *Payee's Name.* Write the name of the person to whom the check is written in the space after "Pay to the order of."

- *Amount in numbers.* Write the dollar value close to the dollar sign. Change should be written after a decimal point, for example, $29.95.

- *Amount in words.* This is written on the line below the payee's name. Begin at the extreme left of the line. Start with a capital letter, and write out the dollar value of the check. To show change, use the word *and,* then the value over 100; for example, "Twenty-nine and $\frac{95}{100}$." If any space remains to the right, draw a line up to the words or numbers so that no blank space is left. For example:

$$\text{"Twenty-nine and } \frac{95}{100} \text{———"}$$

- *Memo.* This is optional. You may write what the check is for in this space.

- *Signature.* This is where you write your name. Do not print.

John E. and Jane Doe   101

Jan. 16, 20 05

PAY
TO THE
ORDER OF ___ABC Publications___ | $29.95

Twenty-nine and 95/100 ——————————————— DOLLARS

Strongbox Savings
South River Office
South River, NJ 08882

FOR ___Mag. Sub.___   John E. Doe

How to Endorse (or Cash) a Check

Endorsing a check allows you to cash, deposit, or transfer to someone else a check made payable to you.

▶ To endorse a check payable to you, turn the check over. On the back at the top, write your name as it appears on the check. (Usually the words "Endorse here" are printed in this space.)
▶ Include your checking or savings account number below your name.

LIST 232

How to Complete a Check Register

You can record to whom you write your checks in a check register or on a check stub. Some check stubs are included at the left of the check, while others are not actually stubs but are part of a register that contains a record of checks written, fees, deposits, and interest (if any) earned. Although check registers and check stubs may vary somewhat in format, they all provide a place to keep accurate records about your checks, and they all include the same general information.

The Check Register

Fill in the following information in the check register:

▶ *Check number.* Write the number of the check in this space.

▶ *Date.* Write the date the check was written in this space.

▶ *To.* Write the name of the person, company, or organization to whom the check was written in this space.

▶ *For.* Write what the check was for here.

▶ *Amount.* Write the amount of the check in this space. To keep a running tally of how much money is in your account, subtract the current check from the previous balance.

Bank Fees

Fees vary from bank to bank. Some banks charge the following fees on checking accounts:

▶ Per-check fee

▶ Monthly service fee

▶ A charge for new checks

▶ A penalty fee if your balance drops below a specific amount

▶ A fee for the use of an ATM (automatic teller machine)

▶ A fee for a bounced check (a check that you wrote but for which you do not have enough money in your account to pay)

Bank service fees are subtracted from the balance in your account. Subtract these fees from the balance in your register.

(Continued)

Deposits

A *deposit* is an amount of money that you put into your account.

▶ When you make a deposit, record the date and the amount of the deposit in the check register.

▶ Add the amount of the deposit to the previous balance.

▶ Keep the receipts of your deposits for your records. Even banks sometimes make mistakes, and the receipt is your proof.

Interest-Bearing Checking Accounts

Some checking accounts pay interest on the money you have in your account. These accounts usually require that you maintain a minimum balance in your account. The interest will be recorded on the bank statement you receive each month. Add the interest to your previous balance in your check register.

LIST 233

How to Balance a Checkbook

A statement is a summary of the transactions made on your checking account. It contains a listing of all the checks that have been cashed, deposits you have made, funds you have withdrawn through an ATM (automatic teller machine), and fees you have been charged. Most banks send a statement each month and also provide electronic services where their customers can view their account transactions on-line. It is a good practice to balance your account according to your check register or check stubs and compare it to the bank's statement. Sometimes either you or the bank may make an error.

Getting Organized

▶ Checks.
  ▪ If your bank sends canceled checks (these are checks that have been cashed and honored by your bank) with your statement, arrange the checks in numerical order. Some banks no longer send canceled checks; instead, they send photocopies of the checks or a mere listing of them. Should you need proof of a canceled check, you can contact the bank.
  ▪ Use your register to mark each check that has been cashed and honored by the bank.
  ▪ Be sure the amount on the check matches the amount of the check on the statement, as well as the amount of the check recorded in the check register.
  ▪ On the bank statement or a separate sheet of paper, list any checks that have not been returned.
  ▪ If your bank charges a fee for each check you write, multiply the fee by the number of checks returned and deduct this amount from your balance.
▶ Be sure all fees and ATM transactions are recorded.
▶ Check your register and statement to make sure that all deposits appear on both.

Finding Errors

You are a rare individual if your check register always matches the bank's statement of your account. If your register and bank statement do not agree, do the following:

▶ Review your account. Has it been credited correctly? This is where your deposit receipts are important.
▶ Review your account and make sure it has been debited correctly. Review the check amounts, fees, and ATM withdrawals.
▶ Redo your addition and subtraction.
▶ Make certain that you have copied the balance correctly to start a new page in your register.

(Continued)

If you made the mistake, recheck everything from that point forward. One error will throw off the rest of the balance. If the bank made the mistake, notify the bank. You may need to provide copies of deposit receipts or cashed checks to prove that you are right.

A Few Suggestions for Doing the Math

► Start with the balance shown on the statement.

► Add deposits made since the last statement.

► Subtract the cashed checks.

► Do not subtract any outstanding checks from the statement period that have not been cashed. Note that these will not yet be debited from the statement.

► Do not subtract any checks you have written since the time of the statement. Note that these will not yet be debited.

► Subtract any check fees.

► Subtract any ATM withdrawals and fees.

► Add any interest. (Some checking accounts offer interest.)

► The balance on the statement should equal the balance in your register.

LIST 234

Loans and Interest

Unless you borrow money from a generous friend or relative, you will have to repay the loan with interest. Note the facts about loans and interest.

▶ A *loan* is a sum of money that is lent at a specific rate of interest. When you borrow money, the money you borrowed becomes a loan.

▶ *Interest* is the cost to you for using someone else's money. It is a fee that banks, for example, charge for lending money.

▶ Interest rates vary based on the economy and the type of loan. Generally the longer you take a loan, the more interest you will pay.

▶ Loans can be repaid over time. For small purchases, a loan might last a year or two. When you buy a car, you might take a loan for between one and five years, but when you buy a house, your loan (called a *mortgage*) might be for 15, 20, 25, or 30 years.

▶ The typical loan is repaid in monthly installments. As you repay the loan, you will be paying off the amount of money you borrowed, plus the interest. (When you borrow money that is to be repaid with interest, you always pay back more than you borrowed.)

▶ Since interest is calculated on the amount of money you borrow, it is helpful to put down as much of your own money as you can when buying something you must finance. This will reduce the amount of the loan. The lower the amount of the loan, the less total interest you will pay (given the same interest rate and time.)

The following formula may be used to calculate the monthly payment necessary to repay borrowed money at interest:

$$M = \frac{A\left(\frac{r}{12}\right)\left(1 + \frac{r}{12}\right)^n}{\left(1 + \frac{r}{12}\right)^n - 1}$$

where M = monthly payment, A = amount borrowed, r = annual rate of interest, and n = the number of months of the loan.

Steps to Making a Budget

A *budget* is a plan that helps you to keep track of your money. A good budget contains your sources of income and your expenses.

1. Decide if you need a weekly or monthly budget. If you choose a weekly budget, you will work with income and expenses for the week. If you choose a monthly budget, your income and expenses will be calculated over a four-week period. For most people, a monthly budget is more practical.

2. Start with a sheet of paper. Divide it into two columns, and label the left side "Income" and the right side "Expenses."

3. List all of your sources of income. Include money you earn from jobs, allowances, and gifts. (If a source of income varies or if you are unsure about it, underestimate its amount.)

4. Add up the total amount of money you receive.

5. List all of your expenses.

   - Include all fixed expenses such as bills, the cost of food, lunches, clothing, entertainment, recreation, and others. (Some people like to include money committed to savings as an expense. This way they are certain to budget for it.)

   - Estimate miscellaneous expenses for sports equipment, special events, gifts, replacement parts for equipment that might break down and need repair, and so forth.

6. Add up all of your expenses, and subtract your total expenses from your total income.

   - If your income equals your expenses, your budget is *balanced.*

   - If your income exceeds your expenses, you have a *budget surplus.* You may save the extra money or keep it handy for unexpected expenses.

   - If your expenses exceed your income, you must spend less or find new sources of income.

LIST 236

Items on Sale

Everybody likes to buy things on sale. Sales always give you a good deal. Or do they? Sale signs or notices may advertise the sale by telling you the dollar savings off the original price of an item, or they may offer only the percent savings. In this case, you will need to do some fast figuring to see how much you will really save.

The following list offers some important words you will need to know if you are to understand the actual value of a sale, as well as some formulas you can use to calculate how good your buy actually is.

Regular price (or marked price)—the price of an item when it is not on sale.

Sale price—the price of an item that is on sale.

Discount—the dollar amount saved (off the regular price).

Rate of discount (percent off)—the fraction of the original price that is saved when an item is bought on sale.

Common Formulas for Discount and Sale Price

Discount = Regular Price – Sale Price

or

Discount = Rate of Discount × Regular Price

Sale Price = (1 – Rate of Discount) × Regular Price

It is also possible to calculate the values of any of the four terms above if you are given two other values. The following abbreviations are used in the formulas on the next sheet:

R = regular price

S = sale price

D = discount

P = rate of discount (expressed as a fraction or decimal)

(Continued)

To Find	Given	Use
Sale price	Regular price and discount	$S = R - D$
	Regular price and rate of discount	$S = (1 - P)R$
	Rate of discount and discount	$S = (D \div P) - D$
Discount	Regular price and sale price	$D = R - S$
	Rate of discount and regular price	$D = PR$
	Rate of discount and sale price	$D = [S \div (1 - P)] - S$
Rate of discount	Regular price and sale price	$P = 1 - \dfrac{S}{R}$
	Regular price and discount	$P = 1 - \dfrac{R - D}{R}$
	Sale price and discount	$P = 1 - \dfrac{S}{S + D}$
Regular price	Sale price and discount	$R = S + D$
	Sale price and rate of discount	$R = S \div (1 - P)$
	Discount and rate of discount	$R = D \div P$

Here are some shortcuts to estimate the discount. Note that these apply only to percents that can be reduced to fractions with numerators of 1.

Example: 25% off the original price of $29.95. Round $29.95 to $30. Since 25% = $\frac{1}{4}$, divide $30 by 4, which equals $7.50. Your savings would be about $7.50. Additional examples of shortcuts:

▶ 50% off: 50% = $\frac{1}{2}$.   Divide the original price by 2.

▶ 33$\frac{1}{3}$% off: 33$\frac{1}{3}$% = $\frac{1}{3}$.   Divide the original price by 3.

▶ 20% off: 20% = $\frac{1}{5}$.   Divide the original price by 5.

▶ 10% off: 10% = $\frac{1}{10}$.   Divide the original price by 10.

▶ 5% off: 5% = $\frac{1}{20}$.   Divide the original price by 20.

LIST 237

How Much to Tip

When others render a service—for example, a waiter taking an order and bringing food—the customer tips him or her. For many people in the service industry, tips are an important source of their income. The standard tip for most services is 15% of the bill. If you have received superior service, you might tip up to 20%, and if you are dissatisfied, you might leave only 10%. The following list shows common tip amounts.

- *Barber:* 15% of the bill.
- *Bartender:* 15% to 20% of the bill.
- *Hairdresser:* 15% to 20% for the stylist; $2 to $3 for the person who shampoos hair.
- *Bellhop:* $1 per bag; an extra dollar for opening the door to your room.
- *Bus tour guide:* $2 to $3 per half-day.
- *Caterer:* 15% to 20% of the bill.
- *Chambermaid:* $2 to $10 per day, depending on the quality of the work and the consideration she shows.
- *Coat check:* $1 per coat.
- *Doorman at a hotel:* $1 for each bag carried into the hotel for you.
- *Driver of hired limousine:* 10% to 15% of the bill.
- *Flower delivery person:* $1 to $10, depending on the amount of the flowers.
- *Furniture delivery person:* $5 to $10 per person or more depending on how many pieces and the weight.
- *Manicurist:* 15% to 20% of the cost for the manicure.
- *Parking attendant:* $1 to $2 for parking your car.
- *Pedicurist:* 15% to 20% of the fee.
- *Pet groomer:* 15% of the cost for grooming your pet.
- *Pizza delivery person:* $2 to $5.
- *Restroom attendant:* $1 for handing you a towel.
- *Taxi driver:* 15% to 20% of the fare.
- *Valet at a hotel:* $3 to $5 for performing special services for you.
- *Waiter or waitress:* 15% to 20% of the bill.
- *Wine steward:* 10% of the wine check.

In general, consider a tip of 15% to 20%, but use common sense.

Copyright © 2005 by Judith A. Muschla and Gary Robert Muschla

LIST 238

Stock Market Words

The stock market is the foundation of America's economy. It is the arena where stocks (shares of ownership in a company) and bonds (long-term debts) are bought and sold. For businesses, the stock market is a place where they can raise money to operate and expand their enterprises. For people who wish to invest money, the stock market offers a chance to buy shares in a company and benefit from the company's growth. While the following list does not contain all the words related to the stock market, it does contain the words that will start you on the road to becoming a savvy investor.

American Stock Exchange (Amex)—the second largest stock exchange in the United States after the New York Stock Exchange.

Bear—an investor who expects stock prices to fall.

Bear market—a period of generally falling stock prices.

Bid—the price a dealer says he or she will pay for stocks or bonds at a given time.

Blue chips—stocks of companies that are industry leaders.

Board of Governors—the officials of the New York Stock Exchange who oversee the trading of stocks and bonds.

Bond—a certificate, written and sold by the government or a company, that promises to repay, with interest, borrowed money.

Bondholder—an individual or institution that owns a bond or bonds.

Broker—a person who buys and sells stocks and bonds for investors.

Bull—an investor who expects stock prices to rise.

Bull market—a period of generally rising stock prices.

Call—an option to buy stock at a preset price until a given day.

Chief executive office (CEO)—the executive with the greatest decision-making authority in a company.

Commission—the fee a stockbroker receives from a customer for buying or selling stock.

Common stock—the typical method by which a company divides its ownership. Owners of common stock have the right to vote at stockholders' meetings and receive dividends (if the company distributes dividends after payments are made to any preferred stockholders and bondholders).

Corporation—a group of people who obtain a charter or certificate of incorporation, which gives them specific legal rights in the operation of their business.

Day trader—an investor who engages in quick on-line trades, hoping to take advantage of small changes in prices.

Director—the principal official of a company; he or she is elected by the stockholders to oversee the management of the company. Directors choose the officers who manage the company.

Dividend—money a corporation pays to its stockholders.

LIST 238

(Continued)

Dow-Jones Average—the average price of 65 important stocks traded on the New York Stock Exchange. Investors often judge the general performance of the overall market based on the Dow-Jones Average.

Earnings—the profits of a company.

Electronic trading—buying and selling stocks and bonds using the Internet; also called e-trading.

Floor brokers—brokers who trade stocks and bonds on the exchange floor for member firms; also called *commission brokers*.

Go public—the action of a private company deciding to sell stock to the public.

Incorporate—the act of starting a corporation.

Institutional investors—organizations that invest money that others have entrusted to them.

Interest—payment for the use of money.

Investor—an individual who buys stocks or bonds with the hope of earning a profit.

Margin—an investment method in which an investor buys stocks or bonds partly on credit, borrowing some of the amount due for the purchase from his or her broker.

Market price—the current price of stocks or bonds.

Municipal bonds—bonds issued by units of government other than the federal government. Interest on municipal bonds paid to investors is usually exempt from federal tax.

Mutual funds—companies that invest money entrusted to them by others.

New York Stock Exchange (NYSE)—the world's largest marketplace for securities.

Odd-lot dealers—brokers who specialize in buying and selling odd lots.

Odd lots—a purchase or sale of stock less than the normal unit of trading, usually fewer than 100 shares.

Offer—the price for which a dealer will sell stock at a given time.

Offering—the initial public sale of stock by an underwriter.

Officer—a company official who is elected by the company's board of directors. A company's officers typically include a president, vice president, secretary, and treasurer.

Options—contracts that permit investors to buy or sell stocks at a prearranged price during a given time period.

Over-the-counter market—stocks that are not traded on a stock exchange.

Partnership—a company owned by two or more people.

Portfolio—the total list of stocks and bonds owned by an investor.

Preferred stock—stock on which dividends must be paid first, before dividends can be paid to owners of common stock.

LIST 238

(Continued)

Profit—the money earned by a company.

Prospectus—a booklet that contains information about a company that issues stocks and bonds to the public.

Quote—the current price of a stock or bond.

Rate of return—the percentage of interest or dividends earned on money that is invested.

Regional exchange—a stock exchange located at a place other than New York City.

Registered representative—the representative of a brokerage house who offers financial advice to customers and buys and sells stocks and bonds for them.

Seat—membership in a stock exchange.

Securities—stocks, bonds, and notes bought and sold on the stock exchanges.

Securities and Exchange Commission (SEC)—the agency of the U.S. federal government that is responsible for regulating the buying and selling of securities.

Share—any of the equal parts into which the stock of a company is divided.

Single proprietorship—a business owned by one person.

Speculator—an individual who invests in risky ventures and stocks in hopes of realizing high profits.

Stockbroker—an individual who is a member of the stock exchange and can buy and sell stocks on the exchange.

Stock certificate—a certificate that shows the number of shares an investor owns.

Stockholder—an individual who owns stock in a corporation.

Ticker tape—a ribbon-like tape on which a ticker machine prints stock prices.

Trade—the buying and selling of stocks and bonds.

Underwriter—an investment bank that buys new stocks from a corporation for the purpose of reselling them to brokers who represent investors.

Wall Street—the center of New York City's financial district.

Warrant—an option that allows investors to buy a given number of common shares of stock at a given price during a specific period of time.

Yield—the amount of interest or dividends an investment earns.

LIST 239

Reading a Financial Page

The listing of stock prices on a financial page may look complicated, but it is not as confusing as it may first seem. Although the formats on some stock exchanges and newspapers may vary a bit, most contain the following information.

▶ Read the information from left to right. Remember that the prices of stocks are quoted in whole or fractions of a dollar, such as 23, 23.25, or $23\frac{1}{4}$.

▶ *Hi* gives the highest price of the stock over the past 52 weeks.

▶ *Lo* gives the lowest price of the stock over the previous 52 weeks.

▶ *Stock* or *Name* tells the name of the company. Names are abbreviated.

▶ *Sym* indicates the official symbol of the stock. This is the same symbol that appears on the ticker tape. The symbols *pf* or *pr* tell you if the stock is preferred. If these letters are absent, the stock is common stock. The symbol *Wt* after a company's name means that the quotation is for a warrant. A warrant allows its owner to buy a specific amount of common stock at a preset price for a certain amount of time. The letter *S* means that the stock has been split (or divided) recently, and *N* means that the stock is new, having been issued during the past 52 weeks.

▶ *Div* refers to the cash dividend per share. A dividend is the money from its profits that a company pays to its stockholders. A blank in this column means that the company has not paid dividends.

▶ *Yld%* states the company's yield. This is the stock's return on the stockholder's investment. You can determine the yield of a stock by dividing the dividend by the closing price. When you divide, work your answer out in decimals and round off to hundredths. Change your decimal answer to its percent equivalent; this will be the yield.

▶ *PE* is the P-E ratio (price-earnings ratio). This shows the relationship between the price of the stock and the company's annual earnings. A P/E of 15 means that the price of the stock is 15 times that of the company's earnings per share over the past four quarters. Stocks that are highly sought by investors usually have high P/E ratios; those that are unpopular usually have low ones.

LIST 239

(Continued)

▶ *Vol 100s* provides information on the volume of shares of the stock traded on the previous day. Take the number in the column and multiply it by 100 to find how many shares were traded yesterday. The letter *z* before the number means that you should not multiply by 100, because the number that appears is the total.

▶ *Hi* provides the stock's highest price yesterday. (This appears on the right side of the page. *Hi* for the previous 52 weeks appears on the left.)

▶ *Lo* provides the stock's lowest price yesterday. (This appears on the right side of the page. *Lo* for the previous 52 weeks appears on the left.)

▶ *Close* provides the stock's closing price yesterday.

▶ *Net Chg* or *%Chg* means "net change" and shows the difference between the closing price of the day and the day before. A minus sign signals that the closing price is lower today than it was yesterday; a plus sign signals that it is higher. Bold type on these numbers means that the price of the stock changed 5% or more.

Most stock exchanges and newspapers provide notes explaining the symbols and abbreviations used on their stock listings.

LIST 240

How to Read a Food Label

Reading the labels that accompany the foods you buy can help you select foods for a healthy diet, which may reduce your risks for developing certain diseases. Too much sodium (salt), for instance, causes high blood pressure in some people; in others, too much cholesterol may eventually lead to a heart attack. The following list spells out how to read a food label.

Serving Size—This tells you how much of the food is usually served at one meal. All of the amounts of nutrients, fats, and sodium are calculated for a single serving. If you eat twice as much as the serving size on the label, you must double the values. If you eat about half of the serving size noted, you will need to divide the values by 2.

Calories—Knowing the calories you consume each day can help you to lose or gain weight.

Total Fat—Most people should reduce the amount of fat they consume. Try to select foods in which the number of calories from fat is significantly fewer than the total calories.

Saturated Fat—Saturated fat plays a major role in raising blood cholesterol and your risk of heart disease.

Cholesterol—Cholesterol has been linked to heart disease. For many people, the more cholesterol they have in their blood, the greater their chances are for suffering a heart attack.

Sodium—Sodium is another name for salt. For some people, too much can lead to high blood pressure, which has been linked to heart disease and strokes.

Total Carbohydrate—Carbohydrates are an important source of nutrients for your body. Foods high in carbohydrates are a good choice for many people, although an overabundance of carbohydrates can lead to weight gain.

Protein—Most Americans eat diets high in protein. While protein is important to good health, most of us eat more protein than we need. Most sources of animal protein also contain fat and cholesterol.

Vitamins and Minerals—Vitamins and minerals are essential for good health. By eating a variety of foods, you should be able to achieve 100% of the vitamins and minerals you need each day.

Daily Value—This helps to put the numbers in perspective. The Daily Value is based on people who eat 2,000 calories each day. If you eat more, your personal Daily Value may be higher; if you eat less, it may be lower.

LIST 240

(Continued)

The following list contains key words on food labels:

Fat Free—contains less than 0.5 gram of fat per serving (may also be referred to as "nonfat").

Low Fat—contains 3 or fewer grams of fat per serving.

Reduced Fat—contains at least 25% less fat per serving than the standard item (may also be referred to as "lower fat").

Lean—contains fewer than 10 grams of fat, 4.5 grams of saturated fat, and 95 milligrams of cholesterol per serving.

Extra Lean—contains fewer than 5 grams of total fat, 2 grams of saturated fat, and 95 milligrams of cholesterol per serving.

Light or Lite—contains at least 33% fewer calories or 50% less fat than the traditional portion of the food.

Cholesterol Free—contains fewer than 2 milligrams of cholesterol and 2 grams or fewer of saturated fat per serving.

Nutrition Facts

Serving Size 1/2 cup (114g)
Servings Per Container 4

Amount Per Serving

Calories 90   Calories from Fat 30

	% Daily Value*
Total Fat 3g	5%
Saturated Fat 0g	0%
Cholesterol 0mg	0%
Sodium 300mg	13%
Total Carbohydrate 13g	4%
Dietary Fiber 3g	12%
Sugars 3g	
Protein 3g	

Vitamin A	80%	•	Vitamin C	60%
Calcium	4%	•	Iron	4%

* Percent Daily Values are based on a 2,000 calorie diet. Your daily values may be higher or lower depending on your calorie needs.

	Calories	2,000	2,500
Total Fat	Less than	65g	80g
Sat Fat	Less than	20g	25g
Cholesterol	Less than	300mg	300mg
Sodium	Less than	2,400mg	2,400mg
Total Carbohydrates		300g	375g
Fiber		25g	30g

Calories per gram:
Fat 9   •   Carbohydrate 4   •   Protein 4

More nutrients may be listed on some labels.

LIST 241

Cholesterol Content of Popular Foods

The fat in the diets of Americans has been linked to greater risks for obesity, heart disease, and a variety of cancers. Most researchers believe that the fat in a person's diet should be less than 30% of his or her total calories. For most people, that can be a difficult goal to reach because many of the foods we enjoy, especially processed ones, are high in fat.

While there are different kinds of fat, the one most diets attempt to control is cholesterol. Foods low in cholesterol also tend to be low in saturated fats, which have been implicated in disease. Avoiding foods high in cholesterol generally reduces the overall amount of fat in a person's diet.

As you look through the following list, you will find that fish, poultry, fruits, and vegetables usually have less fat than other foods. It is for this reason that nutritionists urge people to eat more of these foods.

Both calories and the amount of cholesterol in milligrams are provided in the list. A dash means that the amount of cholesterol is negligible.

Food	*Calories*	*Cholesterol (mg)*
MEAT AND DAIRY PRODUCTS		
3 oz haddock	134	63
3 oz halibut	155	52
3 oz baked cod	152	63
3 oz salmon	145	54
3 oz bluefish	132	63
3 oz canned tuna	168	54
4 oz lobster	104	240
4 oz shrimp	104	150
6 oysters	101	500
3 oz crab	88	113
3 oz chicken (no skin)	169	54
3 oz cooked turkey	170	67
3 oz roast turkey	200	67
3 oz lean roast veal	194	81
3 oz veal cutlet	186	81
4 oz roast lamb	256	84
1 egg (boiled)	78	165
1 egg (omelet)	106	165
1 egg (scrambled)	106	165
2 slices of bacon	98	20
1 hot dog	126	55
1 beef hamburger	102	76
3 oz round beef	198	76
3 oz liver	177	270
3 oz sirloin tip roast (fat trimmed)	156	69
3 oz pork tenderloin	142	26
3 oz spareribs (lean)	285	121

(Continued)

Food	Calories	Cholesterol (mg)
2 slices of ham	172	63
2 slices of bologna	124	60
2 slices of salami	165	55
2 sausages	312	67
1 oz American cheese	82	16
1 oz Swiss cheese	107	26
1 oz Parmesan cheese	129	22
1 oz cheddar cheese	104	36
1 tsp butter	36	13
8 oz whole milk	146	11
1 cup buttermilk	81	1
1 cup nonfat milk	88	—
1 oz sour cream	61	12

FRUITS AND JUICES

Food	Calories	Cholesterol (mg)
1 medium apple	87	—
1 peach	46	—
1 pear	63	—
2 plums	57	—
1 tangerine	50	—
½ grapefruit	75	—
1 orange	70	—
½ cantaloupe	60	—
¼ honeydew	42	—
½ cup of cherries	61	—
½ cup of grapes	54	—
½ cup of watermelon	28	—
½ cup of apricots	54	—
½ cup of raspberries	35	—
½ cup of strawberries	28	—
½ cup apple juice	63	—
½ cup grapefruit juice	66	—

LIST 241

(Continued)

Food	*Calories*	*Cholesterol (mg)*
VEGETABLES		
3 celery stalks	9	—
1 pepper	17	—
1 cucumber	6	—
6 mushrooms	7	—
3 asparagus spears	11	—
1 baked potato	139	—
½ cup of mashed potatoes	94	—
1 tomato	22	—
½ cup of tomatoes	23	—
½ cup of tomato juice	25	—
3 lettuce leaves	20	—
½ broccoli	22	—
½ cup of brussels sprouts	31	—
½ cup of beets	34	—
½ cup of carrots	22	—
½ cup of green beans	14	—
½ cup of cabbage	12	—
½ cup of peas	73	—
½ cup of spinach	23	—
BREADS AND CEREALS		
1 slice of white bread	62	—
1 slice of whole wheat	56	—
½ cup of oatmeal	74	—
½ cup of cornflakes	50	—
½ cup of bran flakes	58	—
½ cup of puffed wheat	21	—
½ cup of noodles	54	—
SPREADS AND DRESSINGS		
1 tsp margarine	36	—
1 tbsp jelly	50	—
1 tbsp peanut butter	92	—
1 tbsp mayonnaise	110	—
1 tbsp Italian dressing	69	—
1 tbsp Russian dressing	76	—
1 tbsp French dressing	67	—
1 tbsp blue cheese dressing	77	—

LIST **241**

(Continued)

Food	Calories	Cholesterol (mg)
SNACKS AND DESSERTS		
1 oz milk chocolate	145	6
1 doughnut	210	20
1 doughnut (glazed)	235	21
1 oz potato chips	147	0
1 oz corn chips	155	25
4 chocolate chip cookies	185	18
1 piece of cream pie	450	8
1 piece of cake	110	64
1 cup of ice cream	270	59

The serving sizes in the list are basics. People often eat more than the portions shown. Also, the caloric content and amount of cholesterol in foods may increase because of the way the foods are cooked or served. For example, chicken is considered to be a low-fat, low-cholesterol food. Chicken fried in fat is not. And although a baked potato has a negligible amount of cholesterol, once it is topped with butter, the cholesterol increases significantly.

LIST 242

A Basic Low-Fat Meal Plan

Most doctors agree that a low-fat diet that controls cholesterol and limits the amount of saturated fat is good for a person's health. By following a diet plan like the one below, you will reduce your risks of heart disease and some types of cancer.

Breakfast

1 serving of fruit or juice

1 serving of cereal with nonfat milk

2 servings of bread (with margarine or jelly)

1 serving of coffee, tea, or nonfat milk

Lunch

1 serving of fish or poultry

1 serving of potato

1 serving of cooked vegetable

1 serving of raw vegetable

1 serving of bread (with margarine if preferred)

1 serving of fruit

1 serving of nonfat milk or juice

Dinner

1 serving of meat, fish, or poultry

1 serving of potato, rice, or noodles

1 serving of cooked vegetable

1 serving of raw vegetables (in a salad with light dressing if preferred)

1 serving of bread (with margarine if preferred)

1 serving of coffee, tea, or nonfat milk

1 serving of fruit

Snacks

Any serving of fruits

Any serving of nonfat or low-fat snacks

Note: Many snack foods are high in fats.

LIST 243

Fat and Food

The diets of most Americans have too much fat. The following information offers facts that can help you manage your diet.

Estimates of Daily Fat Needs

Amount of Calories Consumed Daily	Daily Fat Needs	Calories from Fat
1200	less than 40 g	less than 360
1600	less than 53 g	less than 480
2000	less than 67 g	less than 600
2200	less than 73 g	less than 660
2500	less than 83 g	less than 750

Calculating Fat Intake

Most doctors suggest that a person keep his or her daily fat intake to no more than 30% of the total calories he or she eats. To calculate your daily fat intake, with the goal of keeping fat to 30% of the total calories you consume, do the following:

▶ Multiply your daily calories by 0.3.

▶ Divide the answer by 9.

Most doctors suggest keeping saturated fat to no more than 10% of your daily calories. To calculate your daily intake of saturated fat, with the goal of keeping saturated fat to 10% of the total calories you consume, do the following:

▶ Multiply your daily calories by 0.1.

▶ Divide the answer by 9.

Copyright © 2005 by Judith A. Muschla and Gary Robert Muschla

LIST 243
(Continued)

The Fat in Fast Food

Almost everybody likes fast food. What could be more convenient than cruising into a fast-food restaurant, placing your order, and getting a tasty meal at a reasonable price? Unfortunately, most fast foods are loaded with fat. That is one of the reasons they taste so good, but also the major reason you should eat them in moderation. The following list, which provides the calories and total fat in grams, is based on an average of the foods served at various fast-food chains.

Food	*Calories*	*Total Fat (g)*
Regular hamburger	255	9.8
Large hamburger	425	21.5
Regular cheeseburger	310	14.3
Double hamburger	560	32.0
Fish sandwich	400	16.0
Regular hot dog	270	15.0
Large hot dog	518	30.0
Fried chicken (drumstick)	136	8.0
Fried chicken (wing)	151	10.5
Fried chicken (thigh)	276	19.0
Taco	185	8.0
French fries	214	10.0
Pizza	500	14.5

For many people, just one fast-food meal provides them with more than their daily requirement of fat.

LIST 244

Physical Activities and Burning Calories

Physical activity burns calories. But how much depends on the activity and the individual. The following list contains estimates of the amount of calories you would consume each hour for each pound you weigh as you took part in various activities. You will need to compute your caloric expenditures based on your weight.

Your Weight × Time × Calories per Hour per Pound = Total Calories

Here is an example. Suppose you weigh 120 pounds and play tennis (singles) for two hours. Multiply 120 × 2 × 2.9, which equals 696 calories. This means that you expended 696 calories during those two hours.

Activity	*Calories per Hour per Pound*
Ballroom dancing	1.6
Chopping wood (ax)	2.3
Cycling	2.5
Dancing	2.6
Gardening	3.2
Golf (walking)	2.3
House cleaning	1.6
Jogging	4.2
Jumping rope	3.8
Mowing the lawn	2.7
Racquetball	4.0
Raking leaves	1.5
Rowing machine	3.1
Shoveling snow	3.9
Skating	2.6
Skiing (cross-country)	3.7
Skiing (downhill)	2.5
Soccer	3.7
Swimming	3.8
Tennis (doubles)	1.8
Tennis (singles)	2.9
Volleyball	2.2
Walking (briskly)	2.4
Weight training	1.9

LIST **245**

Finding Your Exercise Heart Rate

While exercise is important for good health, overdoing exercise can be dangerous. One of the best ways to find out if you are working out too much is to know the safe levels of your heart rate during exercise, commonly referred to as your *exercise heart rate range*. It is based on the number of beats your heart makes in 1 minute.

If you have been a couch potato much of your life, are overweight, or suffer from a chronic illness, consult your doctor before undertaking any exercise program. When you begin an exercise program, start slowly, and gradually build up your endurance and strength.

The *American College of Sports Medicine* recommends that you find both 55% and 90% of your maximum safe heart rate. These numbers will serve as the low and high points of your exercise range.

You can find your safe heart rate range by doing the following:

1. Start with 220, which is the maximum recommended heart rate.

2. Subtract your age. This gives you your predicted safe maximum heart rate.

3. To find 55% of your safe maximum heart rate, multiply your safe maximum heart rate (Step 2) by 0.55. This answer will be the low point of your exercise heart rate range.

4. To find 90% of your safe maximum heart rate, multiply your safe maximum heart rate (Step 2) by 0.9. This answer will be the high point of your exercise heart rate range.

5. During exercise or immediately afterward, find your heart rate by taking your pulse. Here is how:

   ▶ Take your first two fingers (not your thumb), and press lightly on the carotid artery just beneath your chin. It is almost in a straight line down from the corner of your eye. Or you may use the pulse close to your thumb or wrist if you wish.

   ▶ Count the number of beats for 10 seconds.

   ▶ Multiply by 6, which will adjust the number of heart beats for a minute. Your pulse should be within the high and low points of your range. If it is too high, slow down; if it is too low and you are not winded or tired, work a little harder.

   ▶ Periodically check your pulse rate during your workouts. Pay close attention to signs that you are overworking. Symptoms such as dizziness, faintness, having trouble catching your breath, excessive sweating, or a pounding in your chest are your body's way of telling you to slow down.

The Numbers in Popular Sports

We often take numbers for granted, seldom realizing how important they are to our lives. Consider sports. Few sports could be played and enjoyed without numbers. How would the score be kept? How could statistics be compiled? How could field sizes be made standard? The following examples show just how important numbers are to many of the sports we watch and play. (For fun, see how many math words and numbers you can find in the following list.)

Baseball

► The game is played on a level field. The infield is called a diamond, but it is really a square with sides 90 feet long. Starting with home plate, at each corner is a base. Each corner is a 90° angle.

► The outfield is flat. Foul lines extend from the first and third baselines to foul poles. Within these lines is fair territory; outside the lines is foul territory.

► The distance from home plate to the outfield varies according to park. The area in foul territory also varies. When the distance to the outfield wall is short, the field is sometimes called a "hitter's park" because it is easier to hit home runs there than in other parks. When the distance is long and there is plenty of foul territory, the field is a "pitcher's park." Balls that would be homers in other parks become long fly outs, and fielders have lots of room to catch foul pops.

► In a standard game, there are 9 fielders. Rosters vary according to the league. Professional teams usually carry 25 players.

► The standard baseball is between 9 and $9\frac{1}{4}$ inches in circumference and weighs between 5 and $5\frac{1}{4}$ ounces. Bats are long and rounded, made of wood or aluminum. (Only wooden bats may be used in the big leagues.)

► Professional games are 9 innings, with 6 outs per inning, 3 for each side. The team that scores the most runs wins.

Football

► American football is played on a field 120 yards long, including the end zones, and 53 yards wide. It is 100 yards from goal line to goal line. The field is divided every five yards by a line.

► At the end lines are goal posts. They may be two parallel uprights extending at least 20 feet from the ground and connected by a crossbar, or as in professional football, there may be one upright post, 10 feet high and topped by a horizontal crossbar from the ends of which extend two vertical uprights.

► Points are tallied in several ways: touchdowns, 7 points; extra points after touchdowns, 1 point when the ball is kicked through the uprights, or 2 points when the ball is passed or carried across the goal line; field goals, 3 points; safeties, 2 points.

LIST 246

(Continued)

- ▶ Each team fields 11 players. There is usually a team for offense and a team for defense. Roster sizes vary according to league.

- ▶ Professional and college games are played in four 15-minute quarters.

- ▶ A football is a prolate spheroid. Its circumference is between 28 and $28\frac{1}{2}$ inches at its long axis and $21\frac{1}{4}$ to $21\frac{1}{2}$ inches at its short axis. It weighs between 14 and 15 ounces.

Basketball

- ▶ A basketball court is a rectangle, no more than 94 feet by 50 feet, and no less than 74 feet by 42 feet.

- ▶ The court is divided in half by a center line.

- ▶ The center court circle is used to start the game with a jump ball. The circle is actually two concentric circles. The inside circle has a radius of 2 feet, and the outside circle has a radius of 6 feet.

- ▶ Vertical backboards are at both ends of the court, usually 6 feet by 4 feet. Baskets are attached to the backboards, 10 feet above the floor. Each basket is 18 inches in diameter and consists of a horizontal hoop of cloth or thin-chain mesh.

- ▶ The foul line is 15 feet from the backboard. Players shoot foul shots from behind the line.

- ▶ Each basketball team fields 5 players on the court, although teams may have rosters of 12 or more, depending on the league.

- ▶ In the pros, games are 48 minutes long, divided into four periods of 12 minutes each. In college, games are 40 minutes long, divided into two halves of 20 minutes each.

- ▶ A basketball is a rubber- or leather-covered sphere with a circumference of 30 inches. It weighs 20 to 22 ounces.

- ▶ Points are scored when the ball is put through the opponent's basket. Field goals are worth 2 points, and foul shots count as 1. Long shots from behind the three-point line count as 3 points.

Soccer

- ▶ Soccer, known as football in many parts of the world, is played on a rectangular field not more than 130 yards long by 100 yards wide and not less than 100 yards by 50 yards.

- ▶ A goal is at each end of the field. The goals are made of a pair of upright posts, 8 feet high and 24 feet apart. A horizontal crossbar runs from the top of one pole to the other. Netting backs the goal.

- ▶ A soccer team fields 11 players, although roster sizes vary.

LIST 246
(Continued)

▶ Games are 90 minutes long, divided into two halves of 45 minutes each.

▶ A soccer ball is a sphere 27 to 28 inches in circumference, weighing 14 to 16 ounces.

▶ Points are scored by kicking or "heading" the ball into the opponent's net. Each goal counts as 1 point.

Ice Hockey

▶ Ice hockey is generally considered to be the fastest of all team sports. It is played on natural or artificial ice on an oval rink. The standard rink is 200 feet long by 85 feet wide. It is enclosed by a board wall about 4 feet high.

▶ A goal with a net, 4 feet high and 6 feet wide, is situated 10 feet from each end of the rink.

▶ Two blue lines divide the rink into three equal zones. In the pros, a center line also divides the rink.

▶ Each team puts 6 players on the ice at the same time, although roster sizes vary by league.

▶ Professional games are divided into three 20-minute periods.

▶ The puck is a hard rubber disk, 1 inch thick and 3 inches in diameter. It weighs between 5.5 and 6 ounces.

▶ Players control the puck with wooden sticks. Sticks, from heel to butt, may be no longer than 59.8 inches, and from blade heel to tip 12.6 inches.

▶ Points are scored when the puck goes into the opponent's net.

Field Hockey

▶ Field hockey is usually played on a grassy field, 100 yards long and between 55 and 60 yards wide. The field is divided into four 25-yard zones.

▶ Goals are located at the center of each goal line, at opposite ends of the field. Each goal is 4 yards wide. Goal posts are 7 feet high and are joined by a crossbar at the top.

▶ The ball is about $9\frac{1}{4}$ inches in circumference and weighs no more than 5 ounces.

▶ Players use wooden sticks that weigh between 12 and 18 ounces. The sticks are curved at one end and flattened on the left side.

▶ Games are divided into two halves of 35 minutes each. Teams play with 11 players each, although rosters vary according to league.

▶ Points are scored by hitting the ball into the opponent's net.

LIST 246

(Continued)

Lacrosse

▶ Lacrosse is played on a field 110 yards long, including 15 yards of a clear area behind each goal. Lacrosse fields are between 60 and 70 yards wide.

▶ Goals are located at opposite ends of the field and are formed by two 6-foot poles connected by a 6-foot long crossbar along the top.

▶ The typical men's team plays with 10 players, and the typical women's team plays with 12. Roster sizes vary according to league.

▶ The ball is between 7 and 8 inches in circumference and weighs between 5 and $5\frac{1}{4}$ ounces. A lacrosse stick is between 5 and 6 feet long. It is hooked on top with strings attached to form a net.

▶ A game lasts 60 minutes. Points are scored when the ball is hit into the opponent's net.

Tennis

▶ The standard tennis court is 78 feet long and 27 feet wide. It is expanded to 36 feet in width for doubles play. The court is divided into two equal parts by a net. Each half court is further divided into service courts.

▶ The ball is made of inflated rubber overlaid with a wool composition. It is between $2\frac{1}{2}$ and $2\frac{5}{8}$ inches in diameter, and weighs between 2 and $2\frac{1}{16}$ ounces. The typical tennis racket weighs between 14 and 16 ounces and is made of wood, steel, or aluminum. It has a rounded head, usually strung with gut or nylon.

▶ Scoring in tennis is calculated in a sequence of four points: 15, 30, 40, and game. No points is called *love*. A tie at 40 is called *deuce*. After deuce, a game is won by two points. Points are scored when an opponent fails to return a ball hit by the server.

Racquetball

▶ This fast-moving game is played on an indoor court that is 40 feet long, 20 feet wide, and has front and side walls 20 feet high. The back wall is at least 12 feet high. A shortline, so named because the served ball must cross over it on a fly, is 20 feet from the front wall and divides the court in half. The service line is 15 feet from the front wall.

▶ The hollow, hard rubber ball is between $2\frac{1}{4}$ and $2\frac{1}{2}$ inches in diameter and weighs about 1 ounce. Rackets are slightly smaller than tennis rackets and are usually made of composite metals, steel, or fiberglass. The rounded head is strung tightly with strong nylon.

▶ Points are scored when a player fails to return a shot of the server.

LIST **246**

(Continued)

Volleyball

▶ Volleyball is played on a court 60 feet long by 30 feet wide. The court is divided into two equal parts by a net. Indoor rules call for clearance of at least 26 feet above the entire court.

▶ A volleyball net is 32 feet long, 3 feet wide, and is composed of a 4-inch square mesh of black or dark brown linen. The top of the net is set 8 feet for men, $7\frac{1}{2}$ feet for women, and 7 feet or lower for children.

▶ The volleyball is an inflated sphere between 25 and 27 inches in circumference, and weighs 9 to 10 ounces.

▶ The typical team consists of 6 players, although rosters vary.

▶ Points are scored when the opposing team fails to return a hit by the serving team.

LIST 247

Finding Statistics in Baseball

Numbers and statistics play a greater role in the understanding and enjoyment of baseball than in any other sport. The details of every game are recorded, from the first pitch to the final out. But it is the numbers that describe a player's performance that are most interesting to many fans.

Players are compared by their "stats." Any baseball fan knows the value of a .300 hitter compared to one who bats .190. Likewise, any pitcher whose ERA (earned run average) is over 7.00 is going to win fewer games than the ace whose ERA is under 2.00. Note that most baseball statistics are given in decimal form in hundredths or thousandths. Following are the ways the most important baseball stats are calculated.

A Team's Winning Percentage

This is the percentage of the games a team wins compared to the total number of games it has played. To find a team's winning percentage, divide the total number of wins by the total number of games played. Work the answer out to thousandths.

▶ Here is the math: A team has won 12 and lost 8 games. 12 divided by 20 = .600. Therefore, the team has a winning percentage of .600 (or 60%).

▶ A winning percentage of .600 is often good enough for first place in the league.

Batting Average

A batting average is the measure of how often a batter gets a hit. A batting average is found by dividing the number of hits by the number of official at-bats. Walks, being hit by a pitch, and sacrifices are not counted as official at-bats. Getting on base on a fielder's choice or an error is counted as an at-bat but not a hit.

▶ Here is the math: 1 hit in 4 at-bats. 1 divided by 4 = .250.

▶ An average of .300 is considered very good; an average of .400 is rare. The last player to hit .400 for a season in the big leagues was Ted Williams of the Boston Red Sox in 1941; his average that season was .406.

(Continued)

On-Base Average

The on-base average (also referred to as the on-base percentage) measures how often a batter gets on base. The on-base average is found by dividing the number of times a player reaches base by the number of times he comes to the plate. The on-base average includes walks, being hit by a pitch, and sacrifices, which are considered to be plate appearances.

▶ Here is the math: In 4 plate appearances, a player has 1 hit, 1 walk, 1 sacrifice, and 1 ground-out. He reached base 2 times in 4 appearances; therefore, 2 divided by 4 is .500.

▶ A player's on-base average is almost always higher than his batting average. An on-base average of .400 over the course of a season is quite good.

Slugging Average

The slugging average (also referred to as the slugging percentage) measures a batter's power hitting. It is found by taking the total number of bases the batter reaches, divided by the number of official at-bats. Obviously, the more home runs, triples, and doubles a batter has, the higher his slugging percentage will be.

▶ Here is the math: In 5 at-bats, a batter hits a single and double. The single is worth 1 base and the double is worth 2 bases, giving him 3 total bases. 3 (total bases) divided by 5 (at-bats) = .600 (slugging average).

▶ In the major leagues, a slugging average over .500 is the mark of a power hitter.

RBIs

RBIs (runs batted in) are the total number of runs that score when a batter gets a hit, base on balls, is hit with a pitch, sacrifices (either a bunt or fly ball), grounds out with fewer than two outs, or reaches base on an infield out. If the batter reaches base because of an error, grounds into a double-play or what would have been a double-play except for an error, no RBI is credited. A batter reaching the 100-RBI plateau in a season has had a fine year.

LIST 247

(Continued)

ERA

ERA (earned run average) represents the number of runs given up by a pitcher during a 9-inning game. The ERA does not include runs scored because of an error or the base runner's initially reaching base because of an error. You can find a pitcher's ERA by multiplying the number of earned runs scored by 9, then dividing by the total number of innings pitched. Because innings have three outs, fractions (thirds) play a part in calculating a pitcher's ERA.

▶ Here is the math: A starting pitcher has pitched $125\frac{2}{3}$ innings (125.67 innings) and has given up 35 earned runs. First multiply 35 times 9, which equals 315. Now divide 315 by 125.67, which equals (rounded off to hundredths) 2.51. This pitcher gives up about $2\frac{1}{2}$ earned runs every 9 innings.

▶ An ERA under 3.00 is considered to be good, and an ERA under 2.00 is considered to be outstanding.

Fielding Percentage

The fielding percentage is a measurement of a fielder's ability to make plays cleanly, without errors. It is found by dividing the number of successful fielding plays by the number of opportunities.

▶ Here is the math: A second baseman has 150 fielding chances and makes 10 errors. That means he has fielded 140 balls cleanly. Divide 140 by 150, which equals a fielding percentage of .933.

▶ A fielder whose fielding percentage approaches .900 is nearing a Gold Glove.

LIST **248**

Geometry and Nature

Examples and applications of the different branches of mathematics can be found throughout the world. No branch, however, rivals geometry in its countless representations. The following list offers just some examples of geometry found in nature.

diatoms—circles

snowflakes—hexagons

chambered nautilus (snail)—spirals

tourmaline (a mineral)—internal structure made of triangles

common starfish—pentagons

honeycomb—hexagons

earthworms—cylinders

salt—cubic crystals

diamond—octahedron

quartz—hexagons

sulfur—rhombic prism

morning glory buds—spirals

daisy head—spirals

pine cone scales—spirals

stalks of many flowers—lines or gentle arcs

tree trunks—lines or cylinders

human beings—show remarkable symmetry: two ears, two eyes, two arms, two hands, two legs, two feet, four fingers and one thumb on each hand, five toes on each foot

other animals—symmetry similar to that of humans

stars—gaseous spheres

planets—solid spheres (in most cases)

moons—solid spheres

galaxies—spirals and ellipticals (some are irregular with no specific shape)

orbits of planets—generally circular (with slight variations)

orbits of comets—elliptical or parabolic

LIST **249**

Your Weight on Other Planets

If you want to weigh less, take a trip to the moon and get on a scale. If instead you want to gain a few pounds, rocket over to Jupiter, and you will weigh about 2.5 times what you weigh on Earth. Because of the different masses of the planets, their surface gravity varies from that of the Earth's. Although your body mass remains the same, the force of gravity pulling down on you is different on other planets. The result: your weight would change.

To find how much you would weigh on another planet in our solar system, use this formula:

Your Earth Weight × Surface Gravity of Planet = Your Weight on the Planet

The following list makes the math easy.

Surface Gravity of the Planets (and Our Moon) Compared to Earth

Earth = 1.00

Earth's Moon = 0.17

Mercury = 0.37

Venus = 0.88

Mars = 0.38

Jupiter = 2.51

Saturn = 1.07

Uranus = 0.93

Neptune = 1.23

Pluto = 0.05

A quick question: If you were in space and there was no gravity, how much would you weigh?

Answer: Nothing. You would be weightless, although you would still have the same mass.

Copyright © 2005 by Judith A. Muschla and Gary Robert Muschla

LIST **250**

Some Mathematical Facts About Space

Should we ever meet beings from other worlds, one of the ways we might communicate with them is through mathematics. Unquestionably, mathematics is a universal language. Travelers from distant stars will surely have concepts of direction, distance, and geometric shapes—after all, every star and planet is a sphere—and this would be a starting point on which to begin communication.

The following list offers some interesting math facts about the universe in which we live. First, some information about our solar system.

- ► Because distances even in our solar system are so long, astronomers use a measure called an astronomical unit (AU). It is equivalent to the average distance between the Earth and our sun, about 93,000,000 miles or 150,000,000 kilometers.

- ► Our solar system consists of our sun, the nine planets, their moons, and countless asteroids, comets, and meteoroids.

- ► Our sun, an average star, is a huge, glowing ball of gases. It is about 4.5 billion years old and is expected to shine for at least 4.5 billion more. While at its center temperatures approach 27,000,000°F, the sun's surface is a mere 10,000°F. The sun's diameter is about 109 times that of the Earth's.

- ► The general orbits of the planets, with a little variation, are circular.

- ► A planet's year is based on its revolution around the sun.

- ► A planet's day is based on its rotation, or how fast it spins on its axis.

- ► While the Earth has one moon, some planets have many and others have none.

The following chart provides additional information about the planets.

| | Average Distance from Sun | | | | |
	AU	Miles in Millions	Period of Revolution*	Period of Rotation**	Number of Moons
Mercury	0.4	36.0	88.0 D	58.7 D	0
Venus	0.7	67.2	224.7 D	243.0 D	0
Earth	1.0	92.9	365.3 D	23.9 H	1
Mars	1.5	141.6	687.0 D	24.6 H	2
Jupiter	5.2	483.3	11.9 Y	9.9 H	16
Saturn	9.5	886.2	29.5 Y	10.7 H	18
Uranus	19.2	1,783.0	84.0 Y	7.2 H	15
Neptune	30.1	2,794.0	164.8 Y	17.0 H	8
Pluto	39.5	3,670.0	248.5 Y	6.4 D	1

*The period of revolution is given in Earth time. D = days; Y = years. Thus, Venus circles the sun in a little over a half of the Earth's year. Pluto, being so far from the sun, takes over 248 Earth years to circle the sun.
**The period of rotation is given in Earth time. H = hours; D = days. Some planets spin on their axis faster than others. How fast they spin entirely around determines the length of their days. On the Earth, a day is about 24 hours long. On Venus, which spins very slowly, one day is equivalent to 243 Earth days. (Talk about a long day!)

LIST 250

(Continued)

While we use astronomical units to measure distances within our solar system, the distances beyond Pluto become too great even for that. To measure the distances between the stars, astronomers use *light-years.*

A light-year is the distance that light travels in one year, which is about 6 trillion miles or 9.5 trillion kilometers.

To work out the math, take the speed of light, about 186,282 miles per second, and multiply it by—

60 seconds in a minute, then 60 minutes in an hour, then 24 hours in a day, then 365.25 days in a year.

Here are some additional out-of-this-world facts:

▶ Our solar system is less than one light "day" across.

▶ Proxima Centauri is the star nearest our sun. It is about 4.3 light-years away, about 25 trillion miles. (Proxima Centauri is part of Alpha Centauri, a triple-star system.)

▶ With our existing technology, the fastest spaceship we could build would take about 94,000 years to reach Proxima Centauri.

▶ The North Star, Polaris, is 782 light-years from Earth.

▶ Our galaxy, the Milky Way, is a group of some 400 billion stars. It is a spiral galaxy, moving through space like a gigantic pinwheel.

▶ The Milky Way is one of estimated billions of galaxies in the observable universe. No one has yet seen to the edge of the universe, if there is an edge.

▶ The next nearest spiral galaxy to the Milky Way is Messier 31. It is about 2 million light-years away.

▶ The farthest known galaxies are several billion light-years from Earth.

Conservation Facts

The conservation of precious resources is vital if we are to maintain the Earth for future generations. When facts are given about conservation, they usually include numbers. The following are a sample.

▶ An average running faucet empties between 3 and 5 gallons of water down the drain every minute. Avoid leaving the faucet running while washing your face, brushing your teeth, or shaving.

▶ Washing the dishes with the tap running wastes up to 30 gallons of water.

▶ Fixing a leaky faucet can save as much as 20 gallons of water each day.

▶ Attaching a low-flow aerator to your faucet or showerhead can reduce the flow by up to 50%. Because air is mixed in with the flowing water, you will not notice the decrease in volume. It is estimated that if every American home installed faucet aerators, up to 280 million gallons of water could be saved daily.

▶ Every time most Americans flush their toilets, they use 5 to 7 gallons of water. Nearly 40% of the water used in American homes is "flushed." By installing an inexpensive displacement device, available at most plumbing supply outlets, you can reduce the amount of water your toilet uses for each flush by 1 to 2 gallons.

▶ Nearly 50% of the garbage Americans produce each day is recyclable.

▶ Each week Americans read about 300 million newspapers that are made from about 500,000 trees. A consistent effort of recycling would easily save millions of trees each year.

▶ Using cloth diapers would save over 1 million metric tons of wood pulp each year.

▶ Recycling glass not only reduces the need for additional landfill space, but also reduces the energy costs needed for making new glass by over 30%.

▶ Making aluminum cans from recycled aluminum reduces the energy costs needed for making new cans by 95%.

LIST 252

Some Helpful Formulas

Mathematicians are not the only people who use math formulas in their work. Scientists use them too. Following are some of the most commonly used formulas.

▶ Area of a Circle: $A = \pi r^2$

A = area, r = length of the radius, $\pi \approx 3.14$ or $\frac{22}{7}$

▶ Area of a parallelogram: $A = bh$

A = area, b = length of the base, h = height

▶ Area of a Rectangle: $A = lw$

A = area, l = length of a side, w = width of a side

▶ Area of a Square: $A = s^2$

A = area, s = side

▶ Area of a Trapezoid: $A = \frac{1}{2} h (b_1 + b_2)$

A = area, h = height, b_1 = length of base 1, b_2 = length of base 2

▶ Area of a Triangle: $A = \frac{1}{2} bh$

A = area, b = length of the base, h = height

▶ Circumference of a Circle: $C = \pi d$

C = circumference, $\pi \approx 3.14$ or $\frac{22}{7}$, d = diameter

▶ Converting to Celsius from Fahrenheit: $C = \frac{5}{9} (F - 32°)$

C = Celsius, F = Fahrenheit

▶ Converting to Fahrenheit from Celsius: $F = \frac{9}{5} C + 32°$

F = Fahrenheit, C = Celsius

▶ Converting to Kelvin from Celsius: $K = C + 273.15°$

K = Kelvin, C = Celsius

▶ Density: $D = \frac{M}{V}$

D = density, M = mass, V = volume

▶ Distance: $d = rt$

d = distance, r = rate at which you travel, t = time spent traveling

LIST 252

(Continued)

▶ Effort: $E = \dfrac{R}{M.A.}$

E = effort, R = resistance, $M.A.$ = mechanical advantage

▶ Energy: $E = mc^2$

E = energy, m = mass, c = speed of light

▶ Force: $F = MA$

F = force, M = mass, A = acceleration

▶ Mean: $\bar{x} = \dfrac{\sum x}{n}$

$\bar{x}$ = mean, $\sum x$ = the sum of the individual numbers, n = the total numbers.

▶ Pressure: $P = \dfrac{F}{A}$

P = pressure, F = force, A = area

▶ Speed: $S = \dfrac{D}{T}$

S = speed, D = distance traveled, T = time

▶ Surface Area of a Cone: $S = \pi r^2 + \pi rs$

S = surface area, $\pi \approx 3.14$ or $\dfrac{22}{7}$, r = length of the radius, s = slant height

▶ Surface Area of a Cylinder: $S = 2\pi r(r + h)$

S = surface area, $\pi \approx 3.14$ or $\dfrac{22}{7}$, r = length of the radius, h = height of the cylinder

▶ Surface Area of a Cube: $S = 6e^2$

S = surface area, e = length of an edge

▶ Surface Area of a Prism: $S = 2B + Ph$

S = surface area, B = area of the base, P = perimeter of the base, h = height of the prism

▶ Surface Area of a Pyramid: $S = B + \dfrac{1}{2}Ps$

S = surface area, B = area of the base, P = perimeter of the base, s = slant height of the lateral faces

LIST 252

(Continued)

▶ Surface Area of a Sphere: $S = 4\pi r^2$
S = surface area, $\pi \approx 3.14$ or $\frac{22}{7}$, r = length of the radius

▶ Velocity: $V = \frac{1}{2}at^2$
V = velocity, a = acceleration, t = time

▶ Volume of a Cone: $V = \frac{\pi r^2 h}{3}$
V = volume, $\pi \approx 3.14$ or $\frac{22}{7}$, r = length of the radius, h = height

▶ Volume of a Cube: $V = e^3$
V = volume, e = length of an edge

▶ Volume of a Cylinder: $V = \pi r^2 h$
V = volume, $\pi \approx 3.14$ or $\frac{22}{7}$, r = length of the radius, h = height

▶ Volume of a Pyramid: $V = \frac{Bh}{3}$
V = volume, B = area of the base, h = height of the pyramid

▶ Volume of a Rectangular Prism: $V = lwh$
V = volume, l = length, w = width, h = height

▶ Volume of a Sphere: $V = \frac{4\pi r^3}{3}$
V = volume, $\pi \approx 3.14$ or $\frac{22}{7}$, r = length of the radius

▶ Work: $W = FD$
W = work, F = force, D = distance

Section Seven

Potpourri

LIST 253

The History of Mathematics

Mathematics is an evolving discipline. It is constantly expanding. Although the ancient Greeks, for instance, began to develop number theory and made major contributions to geometry and logic, they had little inkling of probability, which was not formalized until the mid-seventeenth century.

The following list is a brief history of mathematics, according to topic and mathematician. Note that most dates are approximations. Also note that because so many people have contributed to mathematics throughout history, this list is quite subjective. We picked those individuals and events that we believe are among the most significant. Undoubtedly, several others could be included here.

3000 B.C.	—Egyptians and Babylonians begin the development of number systems and early bookkeeping. Simple problems of arithmetic and geometry necessary to a civilized society become commonplace.
700 B.C.	—Ancient Greece begins to flourish. Greek thinkers contribute number theory, geometry, and logic to mathematics.
530 B.C.	—Pythagoras, one of the most famous of the Greek mathematicians and philosophers, founds the Pythagorean movement. Through their extensive study of numbers, the Pythagoreans helped to establish the scientific foundation of mathematics.
300 B.C.	—Euclid completes the writing of *Elements*, 13 volumes on mathematics.
100 A.D.	—Astronomers begin to develop trigonometry.
170	—Ptolemy, mathematician and astronomer, declares that the Earth is the center of the universe. This becomes the dominant view until the seventeenth century, when Copernicus claims that the sun is the center of the solar system.
200	—Diophantus, a Greek mathematician, develops algebra. He is considered by many to be the father of that subject.
540	—Aryabhata, Hindu mathematician and astronomer, advances mathematics by solving the quadratic equation.
600	—Geometry, algebra, and numeral systems benefit from Hindu and Arabian thought. The idea of zero is introduced.
775	—The translation of Hindu mathematics to Arabic takes place.
830	—al-Khowârizmî, an Arabian, works with algebra, number systems, and equations.
1100	—Arabic and Hindu mathematics spread throughout Western Europe.
1200	—Leonardo of Pisa (Fibonacci), an Italian, works with equations, sequences, series, and pi.
1478	—The first mathematics books are printed.
1489	—The signs of + and − are introduced.
1500	—Negative numbers and perspectives are introduced.
1525	—Great work is done on solving equations.
1540	—Imaginary numbers are explored.

LIST 253

(Continued)

1557	—The Englishman Robert Record writes the first English algebra text.
1560	—The Italian Girolamo Cardano works with complex numbers, equations, and probability.
1580	—Decimals are introduced.
1614	—John Napier, a Scottish mathematician, creates the first system of logarithms. He is the first to use the decimal point.
1630	—The coordinate system is developed.
1640	—René Descartes, French philosopher, scientist, and mathematician, works on systematizing analytic geometry. He also attempts to classify curves according to the types of equations that produce them.
1645	—Pierre de Fermat, a French mathematician, studies various topics, including coordinates, polygons, probability, and the Pythagorean theorem.
1650	—Blaise Pascal, a French philosopher, physicist, and mathematician, formulates one of the basic theorems of projective geometry, known as Pascal's theorem. He also invents the first adding machine.
1680	—The German Gottfried Wilhelm von Leibniz works on calculus, functions, and infinity.
1700	—Work on determinants is done.
1780	—Complex numbers are explored by several mathematicians.
1795	—The metric system is adopted in France.
1800	—Theories on projection are developed. Number theory is explored.
1810	—Carl Friedrich Gauss, a German mathematician, expands the study of number theory.
1820	—The Russian Nikolai Ivanovich Lobachevsky works on geometry.
1840	—Various mathematicians work with matrices, vectors, and symbolic logic.
1870	—George Cantor studies infinite sets.
1905	—Albert Einstein develops the Theory of Relativity.
1910	—Various mathematicians, including Alfred North Whitehead and Bertrand Russell, study the logical foundations of mathematics.
1920	—Emmy Noether studies abstract algebra.
1945	—Various mathematicians do extensive work on computers.
1947	—John von Neumann publishes his work on game theory.
1976	—Kenneth Appel and Wolfgang Haken find a solution of the Four Color Problem.
1976	—Benoit Mandelbrot coins the term *fractal*.
1995	—Andrew Wiles presents the proof of the most famous problem in mathematics: Fermat's last theorem.

LIST 254

Famous Mathematicians Through History

Throughout history, countless men and women have made great contributions to mathematics. The following list identifies some of them. The list is arranged according to nationality and includes the individual's major areas of study. In some cases, a mathematician may have been born in one country and done his or her major work in another. The list demonstrates that mathematics is truly international.

American

Howard Aiken (1900–1973)—Computers

Vannevar Bush (1890–1974)—Computers

R. Buckminster Fuller (1895–1983)—Polyhedra

Josiah Willard Gibbs (1839–1903)—Vectors

Herman Hollerith (1860–1929)—Computers

Charles F. Richter (1900–1985)—Logarithms

Eli Whitney (1765–1825)—Congruence

Frank Lloyd Wright (1860–1959)—Triangles

Arabian

Jamshid al-Kashî (c. 1430)—Pi

Muhammed ibn Mûsâ al-Khowârizmî (c. 780–850)—Algebra, equations, numeral systems

Austrian

Kurt Gödel (1906–1979)—Mathematical systems

George Joachim (1514–1576)—Algebra, trigonometry

Chinese

Tsu Chung-chi (c. 480)—Pi

An Wang (1920–1990)—Computers

Dutch

Luitzen E. J. Brouwer (1881–1967)—Logic

Maurits Cornelis Escher (1898–1972)—Geometry, polygons, symmetry

Willebrord Snell (1581–1626)—Pi

Adriaen Vlacq (c. 1600–1667)—Logarithms

(Continued)

English

Charles Babbage (1792–1871)—Computers

George Boole (1815–1864)—Algebra, logic, sets

Henry Briggs (1561–1631)—Logarithms

Lord William Brouncker (1620–1684)—Pi

Henry Cavendish (1731–1810)—Algebra, scientific notation

Arthur Cayley (1821–1895)—Algebra

Augustus DeMorgan (1806–1871)—Infinity, logic, sets

Charles L. Dodgson (Lewis Carroll, 1832–1898)—Logic

Sir Arthur Eddington (1882–1944)—Infinity

Edmund Gunter (1581–1626)—Computers, logarithms

Thomas Harriot (1560–1621)—Geometry, inequalities

Lord Kelvin (William Thomson, 1824–1907)—Measurement

Lady Ada Byron Lovelace (1815–1852)—Computers

John Machin (1680–1751)—Pi

Thomas Malthus (1766–1834)—Sequences and series

Sir Isaac Newton (1642–1727)—Algebra, calculus, infinity, logic

William Oughtred (1574–1660)—Computers, logarithms

Bertrand A. Russell (1872–1970)—Logic

William Shanks (1812–1882)—Pi

Brook Taylor (1685–1741)—Algebra

Alan Mathison Turing (1912–1954)—Computers

John Venn (1834–1923)—Sets

John Wallis (1616–1703)—Pi

Alfred North Whitehead (1861–1947)—Logic

Egyptian

Ahmes (c. 1650 B.C.)—Circles, fractions

Flemish

Gerhard Mercator (Kremer, 1512–1594)—Projection

Simon Stevin (1548–1620)—Decimals, exponents, fractions, vectors

LIST 254

(Continued)

French

Jean le Rond d'Alembert (1717–1783)—Complex numbers

Comte George Buffon (1707–1788)—Pi, probability

Jean Buteo (c. 1492–1565)—Exponents

Augustin-Louis Cauchy (1789–1857)—Proofs

Pierre Curie (1859–1906)—Exponents

Gérard Desargues (1591–1661)—Projection

René Descartes (1596–1650)—Coordinates, exponents, polyhedra, sets, signed numbers

Pierre de Fermat (1601–1665)—Coordinates, polygons, probability, Pythagorean theorem

Jean-Baptiste Joseph Fourier (1768–1830)—Trigonometry

Evariste Galois (1811–1832)—Equations

Sophie Germain (1776–1831)—Symmetry

Charles Hermite (1822–1902)—Algebra

Joseph I. Lagrange (1736–1813)—Calculus, measurement

Pierre-Simon Laplace (1749–1827)—Calculus

Marin Mersenne (1588–1648)—Variation

Gaspard Monge (1746–1818)—Projection

Nicole Oresme (1323–1382)—Functions

Blaise Pascal (1623–1662)—Algebra, computers, probability, projection

Jules Henri Pincaré (1854–1912)—Topology

Jean Victor Poncelet (1788–1867)—Projection

Francois Viète (1540–1603)—Decimals, equations, pi, exponents

German

George Cantor (1845–1918)—Functions, infinity, pi, probability, sets

Ludolph van Ceulen (1540–1610)—Pi, rational and irrational numbers

Ernest Chladni (1756–1827)—Symmetry

Zacharias Dase (1824–1861)—Pi

Richard Dedekind (1831–1916)—Infinity

Albrecht Dürer (1471–1528)—Projection

Albert Einstein (1879–1955)—Functions, geometry, lines, infinity, vectors

Gabriel Fahrenheit (1686–1736)—Measurement

LIST 254

(Continued)

Carl Friedrich Gauss (1777–1855)—Complex numbers, geometry, polygons, statistics, topology

Hermann Günther Grassman (1809–1877)—Vectors

Werner Heisenberg (1901–1976)—Measurement

David Hilbert (1862–1943)—Infinity, mathematical systems, sets

Johannes Kepler (1571–1630)—Calculus, conic sections, polyhedra

Johann Heinrich Lambert (1728–1777)—Complex numbers, pi, rational and irrational numbers

Gottfried Wilhelm von Leibniz (1646–1716)—Calculus, logic, computers, functions, infinity, numeral systems

Ferdinand Lindemann (1852–1939)—Pi

August Ferdinand Möbius (1790–1868)—Topology

Johann Müller (Regiomontanus, 1436–1476)—Algebra, trigonometry

Emmy Noether (1882–1935)—Algebra

Georg Friedrich Bernhard Riemann (1826–1866)—Geometry

Christoff Rudolff (c. 1500–1545)—Arithmetic operations

Charles Steinmetz (1865–1923)—Complex numbers

Michael Stifel (c. 1486–1567)—Signed numbers

Greek

Apollonius of Perga (262–190 B.C.)—Conic sections

Archimedes of Syracuse (287–212 B.C.)—Angles, calculus, circles, logarithms, pi, polyhedra

Archytas of Tarentum (c. 400 B.C.)—Geometry

Aristotle (384–322 B.C.)—Geometry, infinity, logic

Diophantus of Alexandria (c. 250)—Algebra, equations

Eratosthenes of Cyrene (c. 284–192 B.C.)—Numbers

Euclid of Alexandria (c. 365–300 B.C.)—Geometry, numbers, planes

Eudoxus (408–355 B.C.)—Estimations, geometry

Hipparchus of Alexandria (c. 180–125 B.C.)—Trigonometry

Hypatia of Alexandria (370–415)—Conic sections

Menelaus of Alexandria (c. 100)—Trigonometry

Pappus (c. 300)—Geometry

Plato (427–347 B.C.)—Polyhedra

Claudius Ptolemy of Alexandria (c. 85–168)—Trigonometry

Pythagoras of Samos (c. 585–507 B.C.)—Pythagorean theorem, triangles, trigonometry, variation

LIST 254
(Continued)

Hindu (Indian)

Aryabhata (476–550)—Pi

Bhaskara (1114–1185)—Pi, Pythagorean theorem

Srinivasa Ramanujan (1887–1920)—Algebra

Hungarian

János Bolyai (1802–1860)—Geometry

John von Neumann (1903–1957)—Computers, game theory

Irish

William Rowan Hamilton (1805–1865)—Algebra, complex numbers, vectors

Italian

Girolamo Cardano (1501–1576)—Complex numbers, equations, probability, signed numbers

Pietro Cataldi (1548–1626)—Exponents

Bonaventura Calvalieri (1598–1647)—Geometry

Lodovico Ferrari (1522–1565)—Equations

Scipione del Ferro (1465–1526)—Equations

Antonio Maria Fior (c. 1515)—Equations

Niccolò Fontana (c. 1499–1557)—Equations

Galileo Galilei (1564–1642)—Algebra, functions, infinity

Leonardo of Pisa (Fibonacci, c. 1170–1250)—Equations, pi, sequences and series

Giuseppe Peano (1858–1932)—Logic

Cubastro Gregorio Ricci (1853–1925)—Vectors

Paolo Ruffini (1765–1822)—Equations

Girolamo Saccheri (1667–1733)—Geometry

Leonardo da Vinci (1452–1519)—Projection, triangles

Japanese

Lady Murasaki (c. 978–1031)—Permutations and combinations

LIST 254
(Continued)

Norwegian

Niels Henrik Abel (1802–1829)—Equations

Caspar Wessel (1745–1818)—Complex numbers

Persian

Omar Khayyám (c. 1048–1123)—Geometry

Polish

Nicholas Copernicus (1473–1543)—Trigonometry

Marie Sklodovska Curie (1867–1934)—Exponents

Russian

Christian Goldbach (1690–1764)—Induction

Sonya Kovalevsky (1850–1891)—Sequences and series

Nikolai Ivanovich Lobachevsky (1793–1856)—Geometry

Andrey Andreyevich Markov (1856–1922)—Probability

Scottish

James Gregory (1638–1675)—Calculus, pi

Colin MacLaurin (1698–1746)—Geometry

J. Clerk Maxwell (1831–1879)—Statistics

John Napier (1550–1617)—Computers, decimals, logarithms

John Playfair (1748–1819)—Geometry

A. Henry Rhind (1833–1863)—Circles, fractions, number systems

James Watt (1736–1819)—Lines

Swedish

Anders Celsius (1701–1744)—Measurement

Swiss

Jean-Robert Argand (1768–1822)—Complex numbers

Joost Bürgi (1552–1632)—Exponents, logarithms

Gabriel Cramer (1704–1752)—Matrices

Leonhard Euler (1707–1783)—Functions, pi, polyhedra, sets, topology, vectors

Marcel Grossman (1878–1936)—Vectors

LIST **255**

Mathematical Quotations

Mathematics can be a demanding discipline. The following words may help to inspire and encourage.

"There is no royal road to geometry."

Euclid

"Mathematics, rightly viewed, possesses not only truth, but supreme beauty."

Bertrand Russell

"Mathematics is the queen of sciences."

Carl Friedrich Gauss

"Geometry is nothing if it be not rigorous."

H.J.S. Smith

"There is no inquiry which is not finally reducible to a question of numbers."

Auguste Comte

"Mathematics is the science which draws necessary conclusions."

Benjamin Pierce

". . . It is written in the language of mathematics, and its characters are triangles, circles, and other geometrical figures. . . ."

Galileo, speaking of the universe

"Wherever there is a number, there is beauty."

Proclus

"Mathematics is the gate and key to science."

Roger Bacon

"It appears to me that if one wants to make progress in mathematics, one should study the masters and not the pupils."

N. H. Abel

"The science of pure mathematics, in its modern developments, may claim to be the most original creation of the human spirit."

Alfred North Whitehead

"God made the integers; all else is the work of man."

Leopold Kronecker

"The profound study of nature is the most fertile source of mathematical discoveries."

Joseph Fourier

"When we cannot use the compass of mathematics or the torch of experience . . . it is certain we cannot take a single step forward."

Voltaire

"I cannot believe that God plays dice with the universe."

Albert Einstein

LIST **256**

Careers in Math

When most students, and many adults, think about potential careers in mathematics, they usually think about professions and jobs that work directly with numbers; math teachers come immediately to mind. Yet many other occupations rely on a sound understanding and application of mathematics, as the following list shows.

Accountant	Environmental Planner
Actuary	Financial Planner
Aerospace Engineer	Fundraiser
Air Traffic Controller	Geologist
Appraiser	Geophysicist
Architect	Insurance Agent
Astronomer	Internal Revenue Agent
Attorney	Market Research Analyst
Auditor	Mathematical Technician
Bank Officer	Math Textbook Editor
Bookkeeper	Medical Laboratory Technician
Budget Officer	Meteorologist
Cartographer	Navigator
Casino Manager	Nuclear Engineer
Chemist	Opinion Researcher
Computer Analyst	Pharmacologist
Computer Programmer	Physicist
Computer Software Developer	Product Manager
Computer Systems Engineer	Professor of Mathematics
Credit Manager	Purchasing Agent
Cryptanalyst	Quality Control Supervisor
Curator	Radar Technician
Data Processing Manager	Real Estate Agent
Demographer	Sales Representative
Director of Vital Statistics	Securities Trader
Draftsman or Draftswoman	Seismologist
Economist	Statistician
Efficiency Expert	Structural Engineer
Electrical Engineer	Surveyor
Electrician	Teacher of Mathematics
Electronics Technician	Technical Illustrator
Engineer	Tool and Die Maker

LIST **257**

World Firsts

The following list contains some first-place rankings of our world's natural wonders.

Highest Mountains by Continent

Mt. Everest, Nepal-Tibet, Asia	29,028 ft. (8,848 m)
Aconcagua, Argentina, South America	22,834 ft. (6,960 m)
Mt. McKinley, Alaska, North America	20,320 ft. (6,194 m)
Kilimanjaro, Tanzania, Africa	19,340 ft. (5,895 m)
Vinson Massif, Antarctica	16,864 ft. (5,140 m)
Mont Blanc, France-Italy, Europe	15,771 ft. (4,807 m)
Kosciusko, Australia	7,310 ft. (2,228 m)

Deepest Parts of the Oceans

Mariana Trench, Pacific, southwest of Guam	35,840 ft. (10,924 m)
Puerto Rico Trench, Atlantic, near Puerto Rico	28,232 ft. (8,605 m)
Java Trench, Indian, near Java	23,376 ft. (7,125 m)
Eurasia Basin, Arctic, near Arctic Circle	17,881 ft. (5,450 m)

Average Depths of Oceans

Pacific	12,925 ft. (3,940 m)
Indian	12,598 ft. (3,840 m)
Atlantic	11,730 ft. (3,575 m)
Arctic	3,407 ft. (1,038 m)

Principal Rivers of the World by Continent

Nile, Africa	4,160 mi. (6,695 km)
Amazon, South America	4,000 mi. (6,437 km)
Yangtze, Asia	3,450 mi. (5,550 km)
Congo, Africa	2,900 mi. (4,666 km)
Mississippi, North America	2,330 mi. (3,750 km)
Murray, Australia	1,609 mi. (2,589 km)

(Continued)

Principal Lakes of the World by Continent

Caspian Sea,* Asia-Europe	143,244 sq. mi. (371,000 sq. km)
Lake Superior, North America	31,700 sq. mi. (82,102 sq. km)
Victoria, Africa	26,828 sq. mi. (69,484 sq. km)
Maracaibo, South America	5,217 sq. mi. (13,512 sq. km)
Eyre, Australia	3,600 sq. mi. (9,324 sq. km)

Notable Waterfalls

Angel Falls, Venezuela	3,212 ft. (979 m)
Tugela, South Africa	2,014 ft. (614 m)
Cuquenan, Venezuela	2,000 ft. (610 m)
King George Falls, Guyana	1,600 ft. (488 m)
Krimmler, Austria	1,312 ft. (400 m)
Takakkaw, Canada	1,200 ft. (366 m)
Silver Strand Falls, USA (Calif.)	1,170 ft. (356 m)
Wollomombi, Australia	1,100 ft. (335 m)

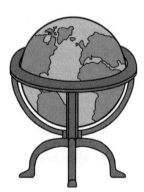

*The Caspian Sea is a lake despite its name. It is entirely bounded by land and fed by several rivers.

LIST 258

Tricky Math Problems

Here are some problems that may tickle your math funny bone. For some answers, your reasoning must also be correct. When you are done, check your total of correct answers against the rating scale.

1. Two U.S. coins total 55¢. One is not a nickel. What are the two coins?

   _____

2. A farmer has 20 cows. He sold all but 12. How many does he have left?

   _____

3. A farmer takes 2 dozen oranges from 3 dozen oranges. How many oranges does he have?

   _____

4. Julius Caesar was given a gold coin dated 48 B.C. He immediately knew it was counterfeit. Why?

   _____

5. A rabbit, chased by a dog, ran from a carrot patch into a nearby woods. How far did it run into the woods?

   _____

6. In the United States we celebrate the 4th of July. Do they have the 4th of July in Canada?

   _____

7. Two students played checkers. Each played 3 games and won 3. There were no draws. Explain how this could be.

   _____

8. Some months have 30 days. Others have 31. How many have 27?

   _____

9. How much is 2 times 4 times 0 times 4 times 2?

   _____

10. Does $\frac{1}{2}$ of 40 = 40 divided by $\frac{1}{2}$?

    _____

LIST **258**

(Continued)

Rating

9–10 correct	Very bright.
7–8 correct	Still bright, but a little dimmer.
5–6 correct	Fading, but okay.
3–4 correct	Getting dark.
1–2 correct	Very dark.
0 correct	It is clear you have been living in a cave.

Answer Key

1. A half-dollar and a nickel. The question said that one coin was not a nickel; the other is.

2. 12

3. 2 dozen

4. The notation B.C. came into use after the birth of Christ, 48 years later.

5. Halfway, because then the rabbit would be running out of the woods.

6. Yes, but Canada does not celebrate it as a holiday.

7. They played against different people.

8. Every month has 27 days.

9. 0

10. No. $\frac{1}{2}$ of 40 = 20; 40 divided by $\frac{1}{2}$ = 80.

LIST 259

Mathematical Palindromes

Most people are familiar with *palindromes* in language: words that are spelled the same forward or backward, like *mom, dad,* and *wow.*

Numbers can be palindromes too. For example, the year 1991 is a palindrome: read from front to back or back to front, it is 1991. The year 2002 is also a palindrome. While a mere list of palindromic numbers might have some interest, the real fun—or frustration—comes in when you work with palindromic sums.

If you take any number, reverse the digits, and add the numbers, continuing this process through as many steps as necessary, you will eventually find a number that is a palindrome. In some cases, you need only a few steps. In others, like the number 89, you need 24 steps before the sum becomes a palindrome of 13 digits.

Following are some numbers for which you will find palindromic sums in just a few steps. We encourage you to try some on your own. (Just make sure the batteries in your calculator are charged!)

One Step: 18

> $18 + 81 = 99$

Three Steps: 257

> $257 + 752 = 1,009$; $1,009 + 9,001 = 10,010$; $10,010 + 01,001 = 11,011$

Four Steps: 372

> $372 + 273 = 645$; $645 + 546 = 1,191$; $1,191 + 1,911 = 3,102$; $3,102 + 2,013 = 5,115$

Seven Steps: 485

> $485 + 584 = 1,069$; $1,069 + 9,601 = 10,670$; $10,670 + 07,601 = 18,271$; $18,271 + 17,281 = 35,552$; $35,552 + 25,553 = 61,105$; $61,105 + 50,116 = 111,221$; $111,221 + 122,111 = 233,332$

Copyright © 2005 by Judith A. Muschla and Gary Robert Muschla

LIST **260**

Roman Numerals

The ancient Romans were master politicians, soldiers, and engineers. They conquered the world from the British Isles to North Africa and Asia Minor, spreading Roman civilization everywhere they went. Rome influenced much of Western civilization that came after it.

Despite all of their achievements, the Romans never developed a practical number system, and once Rome fell, its numbers were supplanted by the Arabic system that we use today. That we still use Roman numerals—on some clocks, watches, movie copyright dates, formal outlines, and decorations—is a tribute to Rome's lasting influence.

Working with Roman numerals is not as hard as it is tedious. You read a Roman numeral from left to right. When a symbol of lesser value comes before a symbol of greater value, subtract the lesser from the greater. For example, IV = 4, or 1 taken from 5. When a symbol of lesser value follows a symbol of greater value, add the two values. Thus, VI = 6, or 5 plus 1. The Romans had no symbol for zero.

Arabic	*Roman*	*Arabic*	*Roman*
1	I	21	XXI
2	II	22	XXII
3	III	23	XXIII
4	IV	24	XXIV
5	V	25	XXV
6	VI	26	XXVI
7	VII	27	XXVII
8	VIII	28	XXVIII
9	IX	29	XXIX
10	X	30	XXX
11	XI	40	XL
12	XII	50	L
13	XIII	60	LX
14	XIV	70	LXX
15	XV	80	LXXX
16	XVI	90	XC
17	XVII	100	C
18	XVIII	200	CC
19	XIX	500	D
20	XX	1,000	M

When they needed large numbers, the Romans sometimes used a bar (called a *vinculum*) over the number to multiply it by 1,000. For example, $\overline{V} = 5,000$, $\overline{X} = 10,000$, $\overline{XV} = 15,000$, $\overline{XXV} = 25,000$, $\overline{L} = 50,000$, $\overline{C} = 100,000$, and $\overline{M} = 1,000,000$.

LIST 260
(Continued)

Some Important Dates in Western History
(as the Romans Would Write Them!)

DCCLIII
753 B.C.
—According to legend, Romulus and his brother, Remus, found Rome. (Of course, there was no B.C. back then.)

CDLXXVI
476
—The Fall of the Roman Empire in the West. The Eastern Empire continued for several more centuries.

MLXVI
1066
—William the Conqueror, duke of Normandy, invades and conquers England.

MCCXV
1215
—The Magna Carta is signed in England, codifying the principle that the king is subject to law.

MCDXCII
1492
—Christopher Columbus discovers the New World.

MDCCLXXXVII
1787
—The U.S. Constitution is written in Philadelphia.

MDCCLXXXIX
1789
—The French Revolution begins.

MDCCCLXI–MDCCCLXV
1861–1865
—The American Civil War.

MCMXIV–MCMXVIII
1914–1918
—World War I.

MCMXXXIX–MCMXLV
1939–1945
—World War II.

MCMLXIII
1963
—President John F. Kennedy is assassinated.

MCMLXIX
1969
—American astronauts Edwin E. Aldrin and Neil A. Armstrong land on the moon.

MCMLXXXIX
1989
—The Berlin Wall comes down.

MMI
2001
—Hijackers seize control of four jetliners. They crash two into the Twin Towers of New York City's World Trade Center, another into the Pentagon in Washington, D.C., and the fourth in Pennsylvania, about 80 miles outside Pittsburgh.

Numbers and Symbolism

Since ancient times, many people have believed that some numbers are symbolic and possess special powers. Many such people dabble in numerology, which is the study of numbers and their influence on life.

1— As the beginning of the counting numbers, 1 is unique. Any number multiplied by 1 remains unchanged; any number divided by 1 remains unchanged too. Geometrically, 1 is related to the point. The number symbolizes beginnings and often is thought to represent balance: male and female, good and bad, light and dark.

2— Unlike 1, the balanced point, 2, like the line that passes through 2 points and extends forever into space in opposite directions, represents contradictions. The ancients believed the numeral 2 symbolized truth and falsehood. We have all heard of second opinions.

3— The number 3 is symbolic of the triangle. In ancient mythology there were 3 Fates, 3 Graces, and 3 Furies. For Christians, 3 symbolizes the Holy Trinity: the Father, Son, and Holy Spirit. Peter denied Christ 3 times. Hindu priests represent their god with 3 heads.

4— The ancients believed the number 4 represented the 4 directions: north, east, south, and west. It also symbolized the ancient elements of earth, air, fire, and water. The 3 sides and 1 base of a pyramid equal 4, which is apparent in the cross and the square as well. The number also represents the 4 basic operations of mathematics: addition, subtraction, multiplication, and division.

5— The ancient Greeks used a 5-pointed star as a secret symbol. Even today, a pentagram is considered by some to hold magical properties.

6— It was believed that God created the universe in 6 days. The number 6 represents 2 triangles and their bases, as well as the cube. It is a number associated with strength and harmony. The Star of David, a 6-pointed star, is a symbol of the state of Israel.

7— The number 7 appears often in the Book of Revelation in the New Testament. During the days preceding the end of the world, 7 plays a prominent role: 7 stars, 7 churches of Asia, 7 spirits before God's throne, 7 horns, 7 vials, 7 plagues, a 7-headed monster, and the lamb with 7 eyes. Scholars have been trying to figure out the meaning of the 7s for centuries. There were also Seven Wonders of the Ancient World, Shakespeare wrote about the 7 ages of man, and "Seventh Heaven" means complete joy. In Japan, according to folklore, there were 7 Gods of Luck.

9— Mystics believed that 9 symbolized wisdom and knowledge. Nine planets revolve around our sun, a fact that some astrologers find significant. The number 9 also possesses regenerative properties. Multiply 9 by any number and the sum of the digits of the answer will be 9. Examples: $9 \times 61 = 549$; $5 + 4 = 9$; $9 + 9 = 18$; $1 + 8 = 9$.

LIST 261

(Continued)

12— Many numerologists believe that the number 12 has special qualities. There are 12 signs of the Zodiac, 12 months in a year, and 12 hours in a day (the other 12 are night). There were also 12 knights of the Round Table. The Bible speaks of the 12 Tribes of Israel, and in the New Testament, there are 12 apostles.

13— In the minds of many, the number 13 is the unluckiest number. Much of the superstition comes from the beginnings of Christianity. The number 13 represents the number present during the Last Supper: Jesus and the 12 apostles.

40— This number appears throughout religion. The deluge reported in the Old Testament lasted 40 days, and another 40 passed before Noah opened the Ark. The Israelites spent 40 years wandering in the wilderness. Jesus spent 40 days and 40 nights in the wilderness, where he was tempted by Satan.

666— According to Christians, this is the Number of the Beast as mentioned in the Book of Revelation. Some scholars believe that 666 referred to Nero, the mad emperor who ruled Rome at about the time the Book of Revelation was written. They believe that the author of the text assigned number values to the Hebrew words that represented Nero, and these equal 666. Perhaps fearing for his head should he be found out by the authorities, the author named Nero, who was certainly considered a beast for the brutality of his rule, by a secret number.

A Number of Coincidences

One of the most fascinating things about numbers are coincidences: events that are linked numerically in some way. Some people are convinced that such occurrences are more than just coincidence; they believe that numbers have special properties and powers at which we can only guess. While that may be stretching reality a bit, after reading the following list, you will need to have a cold, logical mind not to wonder—just a little—if numbers really are nothing more than numerical values.

▶ American presidents Abraham Lincoln and John F. Kennedy were both assassinated. Consider these numbers:

- Lincoln was elected president in 1860 and Kennedy in 1960.

- Both men were assassinated on Friday, the sixth day of the week (assuming the week starts on Sunday).

- Both Lincoln and Kennedy were succeeded by vice presidents named Johnson. Andrew Johnson, Lincoln's successor, was born in 1808 and Kennedy's successor, Lyndon Johnson, in 1908.

- John Wilkes Booth, who assassinated Lincoln, was born in 1839 (according to most sources), and Lee Harvey Oswald, Kennedy's assassin, was born in 1939.

- The names *Lincoln* and *Kennedy* have 7 letters each.

- The names *Andrew Johnson* and *Lyndon Johnson* have 13 letters each.

- The names *John Wilkes Booth* and *Lee Harvey Oswald* have 15 letters each.

▶ Raphael, the famous artist known for painting religious pictures, was born on April 6 and died on April 6. His birthdate and death date were both on Good Friday. Friday is the sixth day of the week.

▶ Harry Houdini, the master magician and escape artist, died on October 31, Halloween.

▶ The square of 13 is 169. Reverse 169, and you get 961. The square root of 961 is 31, which reversed is 13.

▶ Leo Tolstoy, the Russian author, believed 28 to be his special number. He was born on August 28, 1828, his son was born on June 28, and Tolstoy left his home for the final time, before his death, on October 28.

▶ The *Apollo 11* moon flight is related to the number 11. *Moon landing* has 11 letters. President Kennedy initiated the Apollo program, and his last name begins with the letter K, the eleventh letter of the alphabet. The lunar lander touched down on the Sea of Tranquility. *Tranquility* has 11 letters. Neil Armstrong, an astronaut on that flight, was 38 years old; add the digits 3 and 8, and you get 11. Splashdown occurred in the Pacific Ocean, 11 miles from the recovery ship, the aircraft carrier, U.S.S. *Hornet*.

LIST 262

(Continued)

▶ Christopher Columbus discovered the New World in 1492. This opened a new age. In 1942, Enrico Fermi led a team that achieved a nuclear chain reaction. This began the atomic age, a symbolic new world. By switching the middle digits of 1942, you get 1492.

▶ Otto von Bismarck, the German chancellor, was related to the number 3. He served under 3 emperors, had a role in 3 wars, signed 3 peace treaties, owned 3 estates, and fathered 3 children. Even his family crest was related to 3: "in trinity, strength." Trinity means 3.

▶ Three American presidents died on the Fourth of July: John Adams, James Monroe, and Thomas Jefferson.

▶ The composer Richard Wagner was related to the number 13. He was born in 1813. If you add the digits of the year, you will find that they equal 13. His name contains 13 letters. He wrote 13 great works of music. Wagner died on February 13, 1883. Drop the two 8s in the year and you have 13.

Of course, the average, level-headed mathematician knows that these are all coincidences.

Math and Superstition

Over the years people have associated numbers with luck—good or bad. If you are the superstitious type, you need to know the following.

Do These and You Will Have Bad Luck—

▶ Hang a calendar before January 1.

▶ Plant seeds the last three days of March.

▶ Give someone a gift of a knife unless you include a coin.

▶ Start a trip or a project on Friday the 13th.

▶ Break a mirror and suffer seven years of misfortune.

Do These and You Will Have Good Luck—

▶ Find a four-leaf clover.

▶ Carry a silver dollar.

▶ Be the seventh child in a family (although you do not have a choice on this one).

▶ Eat cabbage on New Year's Day, January 1.

▶ Receive a coin with a hole in it.

LIST 264

Mathematical Idioms

Math is so important to our society that it has even found its way into common expressions. There are a lot of mathematical idioms here, and we are sure you can add even more.

Two's company; three's a crowd.

A penny for your thoughts.

Take five.

Don't be square.

Second banana.

Sixth sense.

Ten-four.

Point of no return.

Murder one.

As simple as one, two, three.

Square deal.

Divide and conquer.

Multiply like rabbits.

Another day, another dollar.

An ounce of prevention is worth
    a pound of cure.

Three cheers.

The three R's.

A love triangle.

Give him an inch, and he'll take a mile.

One-way mirror.

Not a second too soon.

Forty winks.

First class.

A bird in the hand is worth two
    in the bush.

Half a loaf is better than none.

A penny saved is a penny earned.

In the top ten.

Two shakes of a lamb's tail.

It takes two to tango.

Give me five.

Looking out for number one.

Twenty-three skidoo.

A third wheel.

Going in circles.

Complete 180.

One-way ticket.

One and only one.

Square meal.

Back to square one.

Feel like you're ten feet tall.

He [she] always gives 110%.

A stitch in time saves nine.

The eleventh hour.

Fair and square.

A ten-gallon hat.

Third rate.

One in a million.

One-way street.

Die a thousand deaths.

Play second fiddle.

Put two and two together.

Don't count your chickens
    before they hatch.

Seventh heaven.

One tree doesn't make a forest.

Two wrongs don't make a right.

Kill two birds with one stone.

Two cents' worth.

LIST 264

(Continued)

Stand on your own two feet.

Two heads are better than one.

A picture is worth
  a thousand words.

Second string.

Dressed to the nines.

One for the book.

One-two punch.

Three strikes and you're out.

On cloud 9.

A catch-22.

My better half.

One-horse town.

Second childhood.

A hot number.

One-track mind.

Baker's dozen.

The odds are a million to one.

Fool me once, shame on you;
  fool me twice, shame on me.

First string.

Third string.

As easy as one, two, three.

One-night stand.

Just like adding two and two.

Batting a thousand.

Fifty cents for one,
  a half-dollar for the other.

Behind the eight ball.

Once in a blue moon.

One-man band.

Second to none.

He [she] is number one.

Seventh son.

A square peg in a round hole.

One-armed bandit.

LIST 265

Numbers in Names and Events

Many names and events contain numbers. Some of them are included in the following list.

Battle of the Three Emperors—in 1805, the Battle of Austerlitz in which Emperor Napoleon of France defeated Francis I of Austria and Alexander I of Russia. During the battle, all three emperors were on the field.

Big Four—after World War II, the United States, Great Britain, France, and the Soviet Union.

Big Three—in 1919, Clemenceau, Lloyd George, and Wilson representing France, Great Britain, and the United States, respectively, at the Paris Peace Conference. In 1945, Churchill, Stalin, and Roosevelt, representing Great Britain, Russia, and the United States, respectively, met at Yalta to discuss the aftermath of World War II.

Chapter 11—a provision of the Federal Bankruptcy Act that permits debtors to transfer the ownership of their assets to new organizations owned by both the debtors and their creditors.

Eightfold Path—in Buddhism: (1) right understanding, (2) right thought, (3) right speech, (4) right action, (5) right livelihood, (6) right effort, (7) right mindfulness, and (8) right concentration.

Fab Four—the 1960s band called The Beatles: John Lennon, Paul McCartney, George Harrison, and Ringo Starr.

Fifth Republic—the government of France since 1958.

First Cause—God as the creator of all things.

First Empire—France under Napoleon, 1804–1814.

First Family—the family of the president of the United States or the family of a state governor.

First Lady—the wife of the president of the United States or the wife of a state governor.

First Republic—France after the overthrow of the monarchy, 1792–1804.

First Triumvirate—in 60 B.C., a coalition formed by Caesar, Pompey, and Crassus.

First World War—the war between the years 1914–1918 (also referred to as the Great War); the first war that involved much of the world.

Five Blessings—in Chinese art, the five bats that are symbols of longevity, wealth, serenity, virtue, and an easy death.

Five Pillars of Islam—the five most important obligations of believers: (1) belief in one God, (2) prayer five times each day facing Mecca, (3) almsgiving, (4) observance of the Ramadan fast, and (5) the *hadj*, which is the pilgrimage to Mecca at least once during a person's life.

Formula One—a race car of specific size, weight, and capacities.

Forty-Niner—someone who took part in the 1849 California gold rush.

4-F—an individual who is rejected for U.S. military service because of physical disability, mental instability, or moral deficiency.

LIST 265

(Continued)

Four-H Club—a club that encourages modern farming practices and good citizenship. The term *Four-H* comes from the idea of improving the "head, heart, hands, and health."

Four Horsemen—in 1924, the backfield of the University of Notre Dame: Harry Stuhldreher (quarterback), Don Miller and Jim Crowley (halfbacks), and Elmer Layden (fullback). Grantland Rice, a sportswriter for the *New York Herald Tribune,* is credited with first using the nickname.

Four Horsemen of the Apocalypse—as prophesied in the Book of Revelation of the New Testament, the personification of war, famine, pestilence, and death.

Four Noble Truths—in Buddhism, (1) life is suffering, (2) the cause of suffering is "birth sin," or craving and desire, (3) only Nirvana can end suffering, and (4) Nirvana can be attained by the Eightfold Path to Righteousness.

Fourth of July—Independence Day in the United States.

Fourth Republic—France's government from 1945 to 1958.

One in Three—in Christianity, a phrase describing the Holy Trinity.

Prime Mover—God.

Second Coalition—in 1799, an alliance of Britain, Austria, and Russia meant to drive France out of Germany, Switzerland, and Italy.

Second Coming—the return of Jesus Christ as prophesied in the New Testament.

Second Empire—France under Napoleon III from 1852 to 1870.

Second Republic—France from 1848 to 1852.

Second World War—the great war that involved most of the world's nations from 1939 to 1945.

Seven Wonders of the Ancient World—the Pyramids of Egypt, the Hanging Gardens of Babylon, Phidias's statue of Zeus at Olympia, the Temple of Artemis at Ephesus, the Mausoleum of Halicarnassus, the Colossus of Rhodes, and the Lighthouse at Alexandria.

Six Day War—from June 5 to June 10, 1967, the war in which Israel defeated Egypt, Jordan, and Syria.

Ten Commandments—the ten obligations God requires of people, as given to Moses on Mount Sinai.

Third Reich—the Nazi dictatorship in Germany from 1933 to 1945.

Third Republic—France from 1870 to 1940.

LIST 265

(Continued)

Third World—many of the countries of Africa, Asia, and South America, often regarded as being underdeveloped.

Third World War—a hypothetical global war, generally assumed to be nuclear.

Thirty Years War—an extensive war between the Catholics of South Germany and the Protestants of North Germany from 1618 to 1648.

Three Stooges (original)—the U.S. comedians Moe Howard, Larry Fine, and Curly Howard.

Three Wise Men—Melchior, Gaspar (also Caspar), and Balthazar, the three kings or Magi who came from the East bearing gifts for the Christ child.

Triple Alliance—from 1882 to 1914, an alliance of Germany, Austria-Hungary, and Italy. In 1717, an alliance of France, the Netherlands, and Great Britain against Spain. In 1668, an alliance of Great Britain, the Netherlands, and Sweden against Spain.

Triple Crown—refers to three horse races: the Kentucky Derby, the Preakness Stakes, and the Belmont Stakes. The horse that wins all three wins the Triple Crown. In baseball, a batter who leads the league in batting average, home runs, and runs batted in is said to win the Triple Crown.

Triple Entente—from 1894 to 1907, an alliance of Great Britain, France, and Russia to counterbalance the Triple Alliance of Germany, Austria-Hungary, and Italy.

Twelve Tribes of Israel—the ancient name of the Israelites.

LIST 266

Math Words Used Every Day

Because many words have multiple meanings, many math words are used in language all the time.

acute	grid	product
area	gross	proper
average	identity	rate
base	imaginary	rational
check	improper	ray
chord	increment	real
complement	index	remainder
complex	interest	rod
composite	intersection	root
cone	interval	scale
coordinates	like	segment
cube	line	series
curve	mean	set
degree	median	similar
difference	mode	simple
double	natural	slope
estimate	null	square
even	obtuse	sum
expression	odd	supplement
extremes	parallel	union
factor	parentheses	unlike
formula	point	variable
fraction	positive	vertex
function	power	volume
grand ($1,000)	prime	whole

LIST 267

Numbers and Language

One of the ways anthropologists and linguists reconstruct the development of a people is through the group's cultural and linguistic heritage. At first glance, English seems as different from Greek as Greek is from French, but a review of the words that represent the numbers from 1 to 10 shows clear linguistic similarities. In fact, the majority of modern European and Indian languages developed from an ancient language known as Indo-European and still show strong linguistic links. In some instances below, the examples are striking.

English	German	French	Spanish	Greek	Latin
one	ein	un	uno	eîs	unus
two	zwei	deux	dos	dúo	duo
three	drei	trois	tres	treîs	tres
four	vier	quatre	cuatro	téssares	quattuor
five	fünf	cinq	cinco	pénte	quinque
six	sechs	six	seis	hex	sex
seven	sieben	sept	siete	heptá	septem
eight	acht	huit	ocho	októ	octo
nine	neun	neuf	nueve	ennéa	novem
ten	zehn	dix	diez	déka	decem

The Greek Alphabet

Many formulas in mathematics use letters of the Greek alphabet as a constant or variable; for example, π, which is equal to approximately 3.14. That we still use these symbols is a tribute to the ancient Greeks, who did so much to further the understanding of mathematics. The names and uppercase and lowercase letters of the Greek alphabet are shown below.

Name	Uppercase	Lowercase
alpha	A	α
beta	B	β
gamma	Γ	γ
delta	Δ	δ
epsilon	E	ϵ
zeta	Z	ζ
eta	H	η
theta	Θ	θ
iota	I	ι
kappa	K	κ
lambda	Λ	λ
mu	M	μ
nu	N	ν
xi	Ξ	ξ
omicron	O	o
pi	Π	π
rho	P	ρ
sigma	Σ	σ
tau	T	τ
upsilon	Y	υ
phi	Φ	φ
chi	X	χ
psi	Ψ	ψ
omega	Ω	ω

Section Eight

Special Reference Lists for Students

LIST **269**

The Math Student's Responsibilities

By accepting the following responsibilities, you will help to ensure your success in math this year.

▶ Come to class on time each day ready to work and learn.

▶ Bring the required materials to class. These might include textbooks, notebooks, binders, pencils, pens, and calculators.

▶ Be curious and inquisitive about numbers. Always look for relationships that will provide insight and understanding.

▶ Recognize that learning the basics is important to any worthwhile activity.

▶ Be diligent in the completion of homework.

▶ Be persistent and determined in your work.

▶ Be willing to try various strategies in solving problems.

▶ Ask questions when you do not understand something.

▶ Accept the challenge of working with classmates in learning the skills and concepts of mathematics.

▶ Be willing to share your understanding of math with others, as well as being open to the ideas of others.

▶ Respect yourself and others in class (and out of class too).

▶ Behave properly, listen to and follow directions, and avoid disturbing others.

▶ Take pride in your work, and never let yourself fall into the trap of believing you cannot do math. Virtually anybody who is willing to work hard enough can succeed.

LIST 270

Overcoming Math Anxiety

By the time they reach middle school, many students are convinced that math is hard. They believe that they do not have the knack for it, and they dread math class. These students worry so much about math that they almost guarantee themselves failure. If you are one of these students, there is much you can do to overcome your math worries with math success.

▶ Realize that boys and girls can do math equally well. The same is true for people of various ethnic groups. There is no one type of person who is predestined for math greatness.

▶ Keep your mind open and your emotions down. If you let yourself think that math is hard or impossible, it will be.

▶ Be persistent in working on your math. Remember that everybody makes mistakes. Learn from your mistakes.

▶ Get in touch with your math feelings. When your mind starts to flood with math worries, stop, take a deep breath, and clear your thoughts. If you are at home, get up and walk around. If you are in class, look out the window for a moment or two. Then redirect yourself to the problem you are working on, and tell yourself that you can do it.

▶ Write your math worries on a sheet of paper—for example, "I can't do word problems." Then write possible solutions—for example, "I can look for key words that will help me to understand the problem, ask my teacher for help, and work on more problems for practice." Seeing your worries in print and ways you can overcome them will help put things in perspective. Every problem has a solution.

▶ Make a commitment to study. Being prepared is one of the best ways to reduce worry.

▶ View overcoming math anxiety as just another problem you can solve.

▶ When you do not understand something, do not be afraid or embarrassed to ask your teacher. Other students probably have the same question.

▶ Keep notes, and review them as necessary.

▶ Study with a friend. The companionship can make the worries easier to handle.

▶ Keep a sense of humor. So you missed a problem. It is not the end of the world. Consider getting a wrong answer as simply one more step to finding the correct answer.

Strategies for Taking Math Tests

When their teacher announces an upcoming math test, many students experience a rush of anxiety. For some, that anxiety can sabotage their performance and lower their scores. The following test-taking strategies can reduce your apprehension and boost your grades.

Before the Test

▶ Prepare by studying the types of problems that might be on the test. If you are given a study guide, be sure you understand the material on it. Practice any examples.

▶ Try to anticipate new or trick problems. Create some examples yourself, and try working them out.

▶ If you are having trouble with some of the material, ask your teacher for help a few days before the test. This will give you time to master the problems.

▶ If you find it helpful, study with a friend. Working with a classmate gives you the chance to share insights and talk about specific problems. It also reduces the feeling of isolation that you are the only one who finds some of the problems difficult.

▶ Think positively about the test. Do not let the comments of others about how hard the test will be affect you. It will not be hard if you are prepared.

▶ Recognize that most people are anxious or nervous about tests. Such feelings usually pass once you start taking the test.

▶ Employ mental imaging. Imagine yourself doing well on the test. Picture yourself solving the problems and earning a high score. This can help build confidence. Professional athletes often use mental imaging to improve their performance.

▶ Prepare yourself physically. Get a restful night's sleep, eat a good breakfast, and arrive at class on time. Rushing to avoid being late will likely upset your concentration for the upcoming test. If your test is in the afternoon, remain calm throughout the day, and eat a good lunch.

▶ When you go to class, be sure you have everything you will need: pencils, erasers, calculator, and so forth. Do not forget your glasses or contacts. Bring plenty of tissues if you have the sniffles.

During the Test

▶ Listen carefully to any instructions. If you have any questions or do not understand something, ask for clarification.

▶ If you feel nervous, take a few deep breaths, and for a few moments, think of a favorite place or an activity you enjoy doing. Getting your mind off the test will help you to regain your composure.

LIST 271

(Continued)

▶ Read directions carefully.

▶ If you are taking a standardized test, do the sample items. Not only are they examples of the problems on the test, but they will help you to warm up.

▶ If you are using a standardized answer sheet, make certain that you are putting the answers in the right places. Use a notecard or your finger to help you stay in the right spot.

▶ Pace yourself. Work quickly but accurately. Make sure that your answers are clear.

▶ Do not waste time on hard problems that have you stumped. Move on to easier ones, but remember to go back. Place a check by the numbers of questions you must return to.

▶ Budget your time. Spend the most time on portions of the test that are worth the most points.

▶ On multiple-choice tests, if you get stuck on an item, eliminate the answers you know cannot be right and work from there. If necessary, make an educated guess from the answers you cannot eliminate.

▶ If time remains after you have completed the test, go over it and double-check your answers.

▶ If you run out of time on a standardized test and if unanswered questions will be marked wrong, *guess*. Fill in an answer for the remaining questions. You have nothing to lose.

▶ For standardized tests, be sure you have filled in all the spaces neatly and darkly and that you have not left any stray marks on the answer sheet.

After the Test

▶ If you did well, reward yourself.

▶ If you did not do as well as you believe you could have, learn from your mistakes, and resolve to do better next time.

Steps for Solving Word Problems

If you find solving word problems to be difficult, the following suggestions can help you be more successful.

► Read the problem carefully. If necessary, read it again. A third reading may even be needed.

► Focus your attention on the question. Identify the problem. What is it really asking? Look for key words like *sum, total, all together, about, difference,* and *how much more* (or *less*) *than.*

► Decide what operation or operations you will need to find the answer.

► Develop a plan to solve the problem. Pick out the facts needed to solve the problem. Ignore unnecessary information.

► Draw a diagram of the problem. This can help you to see relationships.

► Work out the problem accurately.

► Double-check your work.

► Ask yourself if you have answered the question. Is your answer logical? Does it make sense? Are units correct?

LIST **273**

Problem-Solving Strategies
for Middle School Students

It is always easier to solve math problems when you have a plan. Following are some suggestions to help you solve those tough problems.

- ► Read and study the problem slowly and carefully. Be sure you know what it is asking and what you are to find.

- ► Circle the facts needed to solve the problem. Ignore any facts that are not needed.

- ► To understand the problem fully, you might do the following:
  - Draw or sketch a picture.
  - Look for patterns.
  - Write equations.
  - Make a table or chart.
  - Work backward.
  - Try to make the problem simpler; for example, substitute whole numbers for fractions.

- ► Plan what operation or operations you will need to solve the problem.

- ► Work out the problem. Make sure that your math is accurate.

- ► Double-check your work.

- ► Ask yourself if your answer is logical and makes sense.

LIST 274
Problem-Solving Strategies for the Upper Grades

A student once asked her math teacher what was the best way to solve a particular problem. The teacher answered, "The way that works best for you."

Many strategies can be used to solve math problems. When you decide that there are only one or two, you limit your options and reduce the chances of finding the correct answer. Following are some helpful problem-solving strategies.

Understand the Problem

- ▶ Read and study the problem carefully. If necessary, reread the problem.
- ▶ Decide what the question is asking.
- ▶ Find the important information.
- ▶ Eliminate any unnecessary information.
- ▶ Supply any missing facts.

Work Toward a Solution

- ▶ Decide on the operation or operations you will need to use.
- ▶ Use trial and error (sometimes called guess and check).
- ▶ Write equations.
- ▶ Use logic.
- ▶ Design tables or charts to organize data.
- ▶ Sketch or draw a model.
- ▶ Look for patterns and relationships.
- ▶ Make comparisons.
- ▶ Change your point of view. Look at the problem from different angles.
- ▶ Work backward.
- ▶ Recall a similar problem, and use the same strategy.
- ▶ Simplify the problem. Replacing fractions with whole numbers often makes it easier to recognize which operations to use.
- ▶ As you work toward a solution, make notes of your steps in the margin of your paper or on scrap paper. This will help you to keep track of your progress.
- ▶ Periodically review what you have done.
- ▶ Be patient and persistent.

After You Find a Solution, Check Your Work

- ▶ Make sure you used all the important information.
- ▶ Check your calculations.
- ▶ Make sure your answer makes sense. For example, no one pays $2,500 for a parking ticket; insert the decimal point, and $25.00 becomes the logical answer.

LIST 275

Characteristics of Good Problem Solvers

Successful problem solvers share many of the same characteristics, which they use in working out hard problems. How many do you possess? How many can you make a part of your "problem-solving" personality?

Good problem solvers do many of the following:

- ▶ Work actively toward solutions.
- ▶ Search for key ideas.
- ▶ Identify important information.
- ▶ Ignore unimportant information.
- ▶ Supply missing information.
- ▶ Recognize hidden questions.
- ▶ Work carefully.
- ▶ Follow a step-by-step method.
- ▶ Recognize relationships.
- ▶ Identify patterns.
- ▶ Try various strategies.
- ▶ Look at a problem from various angles.
- ▶ Are open to new ideas.
- ▶ Keep notes of their attempts at solutions.
- ▶ Recheck their facts.
- ▶ Use logic.
- ▶ Rely on past experience in solving problems.
- ▶ Are confident that they can solve most problems.
- ▶ Are persistent.

Steps for Writing Word Problems

You are probably familiar with reading and solving word problems. Writing word problems, which you can then share with classmates, is an excellent way to improve your own problem-solving skills. When you construct your own word problems and see what goes into them, you learn what to look for in the problems other people write.

When you are writing word problems, it is helpful to follow the steps of the writing process: prewriting, drafting, revising, editing, and publishing (sharing). You have likely learned about the writing process in your English classes. There are various activities in each stage. You might do some or all of them.

Prewriting

▶ Think of a goal. What do you want others to learn from your problems?

▶ Consider your audience. Who are you writing these problems for? Younger students? Classmates?

▶ Generate ideas for problems.

▶ Gather necessary facts.

▶ Conduct research if necessary.

▶ Analyze your ideas.

▶ Organize your ideas.

▶ Focus your topics.

Drafting

▶ Write the problems.

▶ Rearrange your research as needed.

▶ Elaborate on your ideas.

Revising

▶ Polish your writing.

▶ Rethink or rearrange your information.

▶ Clarify your work.

▶ Rewrite your material where necessary.

▶ Eliminate any unnecessary information or words.

▶ Check facts.

▶ Conduct additional information if necessary.

▶ Seek the opinions of your classmates or peers.

LIST 276

(Continued)

Editing

- ▶ Proofread for accuracy.
- ▶ Make any final revisions.
- ▶ Correct any remaining mistakes in mechanics.

Publishing or Sharing

- ▶ Share your work with classmates.
- ▶ Produce copies of your problems and make them available to others.
- ▶ Publish your word problems with the problems of other students in a class word problem book.
- ▶ Display your problems for others to see and solve.

Math Journal Guidelines
for Students

Keeping a journal can be helpful in math class. You can use a journal to note questions, insights, and reflections you have about your math work. The journal then becomes a record of your progress, showing your growing understanding of math. Following are suggestions for keeping a math journal.

- ▶ Use a standard spiral note, or designate a section of a looseleaf binder for your journal. (Some students prefer to keep their journals on a computer and print out sections as necessary.) If you use a separate notebook and run out of space, simply start a new one and number it (2, 3, 4, and so on).

- ▶ Use your math journal only for math.

- ▶ Date your entries. Some students find that writing the topic at the top of the page is useful if they need to look up that topic later.

- ▶ Be willing to write about math topics and problems that interest you. Journals need not be kept only for notes.

- ▶ Review your journal periodically. You might find that you can now expand on some pieces because you have gained more understanding.

- ▶ Share some of your entries with other members of the class.

- ▶ Review your journal at the end of each marking period to see how your understanding of math is growing.

Some Suggested Math Journal Topics

- ▶ Notes on how to solve certain problems
- ▶ Solutions to similar types of problems
- ▶ Various problem-solving strategies
- ▶ Reflections or impressions about math class
- ▶ Expression of frustration about a tough problem
- ▶ Expression of joy (or relief!) after solving a tough problem
- ▶ Questions (specific or general)
- ▶ Comments about interesting problems
- ▶ Alternate problems
- ▶ Stumbling points on what you found to be confusing or especially difficult
- ▶ Daydreams on math problems
- ▶ A math insight you would like to share with others

LIST 278

Math Web Sites for Students

The Internet offers a truly astonishing amount of information for math students of all ages. The following list and their links provides an assortment of Web sites that promise to keep student mathematicians busy.

- ▶ *Coolmath:* www.coolmath.com
- ▶ *The Math Forum:* http://mathforum.org/students
- ▶ *Allmath:* www.allmath.com
- ▶ *Count On:* www.mathsyear2000.org
- ▶ *Figure This:* www.figurethis.org
- ▶ *NCTM Math Activities:* www.domath.org
- ▶ *Math Stories:* www.mathstories.com
- ▶ *Hotmath:* www.hotmath.org
- ▶ *The World of Math Online:* www.math.com/
- ▶ *Internet Mathematics Library:* http://mathforum.org/library
- ▶ *Mudd Math Fun Facts:* www.math.hmc.edu/funfacts
- ▶ *Webmath:* www.webmath.com
- ▶ *Ask Dr. Math:* http://mathforum.org/dr.math
- ▶ *Math World:* http://mathworld.wolfram.com

To find more Web sites, go to Yahooligans at www.yahooligans.com. Under School Bell, click on Math, and a list of some of the best of current math sites will be displayed.

You can also do a general search. Try "math for students," and you will be rewarded with numerous sites.

For additional information on math and related topics, try the Internet Public Library: www.ipl.org. It has divisions for kids and teens, which can be accessed by clicking on the appropriate link.

Yet another source is *A Maths Dictionary for Kids* by Jenny Eather: www.amaths dictionaryforkids.com.

Section Nine

Lists for Teachers' Reference

LIST 279

The Math Teacher's Responsibilities

Just as math students have responsibilities that will help to ensure their success in math classes, math teachers must accept responsibilities too. To be as effective as he or she can, every math teacher should:

- ▶ Prepare and present lessons that will enable students to learn the math skills and concepts contained in their curriculums.
- ▶ Offer clear directions, explanations, and deadlines for assignments.
- ▶ Support the Standards of the National Council of Teachers of Mathematics (NCTM).
- ▶ Develop and maintain a comfortable and orderly classroom atmosphere that will promote and support learning.
- ▶ Develop a clear set of classroom rules and procedures, and make certain that students understand and follow them.
- ▶ Provide a fair system of grading, and make sure that students understand how their grades are determined.
- ▶ Evaluate student progress regularly, and provide feedback.
- ▶ Encourage students to ask questions, and strive to answer them.
- ▶ Be demanding of work, but also considerate of feelings.
- ▶ Help students learn problem-solving skills.
- ▶ Help students to appreciate the importance of math in their everyday lives.
- ▶ Help students use technology such as calculators and computers.
- ▶ Help prepare students for standardized testing.
- ▶ Be willing to listen to student concerns and problems.
- ▶ Applaud the efforts of students, and take satisfaction in their individual growth.

LIST 280

Upgrading Your Mathematics Curriculum

While teaching some subjects the old-fashioned way is still good enough, this is not true for teaching math. For students to be successful in our global society, they must master high-level problem-solving skills and be able to use technology competently. The following list offers some suggestions you might consider to upgrade your math curriculum and keep it on a par with the Standards of the NCTM.

Grades 4–8

Greater Emphasis

- ► Patterns and relationships
- ► Relevant, real-life problems, in a variety of formats
- ► Open-ended problems
- ► Written explanations of reasoning to find solutions
- ► Group problem solving that emphasizes communication and collaboration
- ► Application of mathematics to other areas
- ► Incorporation of various disciplines in problem solving
- ► Estimation in problem solving
- ► The use of calculators and computers in problem solving
- ► The use of logic in problem solving
- ► Checking for reasonable answers
- ► Interpretation of graphs, tables, charts, and diagrams
- ► Inductive and deductive reasoning
- ► The use of models, diagrams, and sketches in problem solving
- ► Statistical methods

Less Emphasis

- ► One-step word problems
- ► Problems that use irrelevant or meaningless situations that are of little interest to students
- ► Reliance on clue words to determine which operations to use in solving word problems
- ► Drill
- ► Development of skills out of context
- ► Memorization of formulas and facts

LIST 280

(Continued)

Grades 9–12

Greater Emphasis

- Problems that require investigation, either individually or in groups
- Integrated math problems that tie into other disciplines
- Problems designed around real situations
- Written explanations that detail reasoning in the solving of problems
- The use of calculators
- The use of computers, particularly in problem solving and to develop conceptual understanding
- Inductive and deductive reasoning
- Integration of various subjects and the application of different areas of mathematics
- Modeling

Less Emphasis

- One-step word problems
- Drill
- Paper-and-pencil evaluations and solutions
- Memorization
- Paper-and-pencil graphing of solutions

LIST **281**

Materials Every Math Classroom Should Have

It is tough teaching math with just a textbook or workbook. You need lots of materials and supplies. The following list provides the basics.

General Supplies

Calculators	Tape
Pens	Glue
Pencils	Posters
Paper	Poster paper
Colored pencils	Grid paper
Rulers	Erasers
Meter sticks	String
Graph paper	Overhead projector
Scissors	Transparencies
Markers	Markers for transparencies
Compasses	Math dictionary
Protractors	Resource books
Computers (with Internet access)	Printers
Software	

Supplies You Will Need According to Topic

Stopwatch	Base 10 blocks
Cuisenaire® rods	Shape tracers
Decimal squares	Pattern blocks
Fraction squares	Spinners
Fraction circles	Pentominoes
Algebra tiles	Tangrams
Geoboard and rubber bands	Snap cubes
Two-color counters	Color tiles
Mirrors	Dice
Polyhedra dice	Thermometers
3-D geometrical models	Measuring cups
Scales	Balance
Platform spring balance	Liter cube
Graduated cylinder	Mass set
Manipulatives for the overhead projector	Timer

LIST 282

Expanding the Horizons of Your Math Classes

The learning environment of any math class can be enhanced by expanding its horizons and broadening its scope. Consider the following suggestions.

- ▶ Keep your classroom bright and cheerful.
- ▶ Promote and foster fresh ideas, openness, and sharing.
- ▶ Encourage student creativity and discovery.
- ▶ Treat mathematics as a subject that everyone—male, female, and members of all ethnic groups—can learn.
- ▶ Encourage all of your students all of the time.
- ▶ Remember that students are individuals and learn at their own pace.
- ▶ Demand an orderly classroom. Explain your expectations and requirements so that students know what is expected of them.
- ▶ Model appropriate behavior.
- ▶ Move around the classroom and interact with your students. Become an encourager, guide, and nurturer as well as provider of information.
- ▶ Be fair, consistent, and firm in your discipline.
- ▶ Encourage students to consider and explain their reasoning during problem solving.
- ▶ Use group work to foster the sharing of ideas.
- ▶ Incorporate other disciplines in your math class.
- ▶ Encourage writing. Show how math applies to science, social studies, literature, and art whenever possible.
- ▶ Connect math to real-life problems and situations.
- ▶ Show students the relevance of math to everyday living.
- ▶ Encourage the use of calculators and computers.
- ▶ Use manipulative materials.
- ▶ Use fair assessments that reflect the skills that have been taught.
- ▶ Remember that students will usually rise to your expectations. Maintaining high, realistic goals in a positive, nurturing environment will help your students to realize their greatest potential.

The Math Teacher's Management Strategies

The following tips can help you manage your day more effectively.

- ▶ List and prioritize your tasks each day. Do the most important things first.

- ▶ Arrange your tasks according to times that are best for you. For example, if it is easier for you to grade papers at home, do your grading there rather than at school. Use your time in school for other tasks.

- ▶ Try to stagger tests and projects so that you are not overwhelmed with too many papers to correct over one weekend. Get in the habit of taking one class's papers home each night.

- ▶ Try to handle each paper only once.

- ▶ Keep the papers of different classes separated. Use folders or large envelopes. A tote bag is great for carrying things from classroom to classroom and to and from school.

- ▶ Keep your desk organized and your file cabinets in order. Few other things are more aggravating than being unable to find something that is right in front of you but buried beneath the pile on your desk.

- ▶ Remember that everyone has a limit. Do not assume extra tasks or responsibilities just because no one else will take them.

- ▶ Set time limits to complete tasks and activities. Try to stick to them within reason.

- ▶ Start meetings on schedule, and keep them focused on the agenda.

- ▶ Whenever you need to wait (at the dentist's, for example), take along plenty of reading materials. Use this time to catch up.

- ▶ Enlist the help of students whenever possible. Artsy students are often quite happy to do the hallway bulletin board you keep putting off.

- ▶ Encourage students to work together and help each other in solving difficult problems.

- ▶ At the beginning of the year, discuss your classroom rules with students, and explain procedures. When students understand what is expected of them, they are more likely to act and behave appropriately. Once your rules are stated, be sure to enforce them.

- ▶ When making telephone calls, do them all in one sitting.

- ▶ When you cannot reach a parent and must have that parent call you back, offer a time and a telephone number (it is usually best to leave a school number) where you may be reached. This reduces the chances of conducting a conversation through answering machines or secretaries.

- ▶ Use e-mail to communicate brief messages to colleagues, administrators, and parents. It is best to use your e-mail address at school, and not your personal e-mail address, when communicating with parents.

- ▶ Learn to recognize when enough is enough. If you burn out today, tomorrow is lost.

LIST 284

Managing a Successful Math Class: Student Problems and Solutions

You are indeed a rare teacher if your students never experience problems with discipline or motivation or worse! Although the following suggestions are by no means cure-alls, they can help.

***Problem:* The student does not complete assignments.**

Possible solutions: Speak to the student individually about finishing work on time. Perhaps the work is too difficult, and his class needs to be changed. If he remains in your class, monitor his work closely. If there is no improvement, contact the student's parents or guardians, and make arrangements for him to give up a free period or stay after school to complete the work. If necessary, consult a guidance counselor about the student. There may be an underlying problem.

***Problem:* The student is easily distracted.**

Possible solutions: Speak to this student about her behavior, and stress the importance of remaining on task. Position her desk close to yours, and seat her near quiet students who are less likely to distract her. When group work is required, be sure to place her with students who will remain focused on their assignment. Encourage her, and offer praise for appropriate behavior.

***Problem:* The student is disruptive or argumentative.**

Possible solutions: Address the unacceptable behavior quickly. Waiting for it to simply go away usually results in the problem escalating. Explain to the student why such behavior is not acceptable and will not be tolerated. You might try to draw the student out. Ask him why he is misbehaving. An honest question from a caring adult often results in surprising openness. If that approach does not work, you may need to mention the consequences should the behavior continue. If it does, follow the standard disciplinary procedures of your school, which might include detention, a meeting with the vice principal, or a conference with parents or guardians.

***Problem:* Several students are not getting along.**

Possible solutions: The first step is to separate the students. Speak to them as a group, and discuss the nature of the problem and how it might be solved. Offer strategies such as focusing on positive comments and actions, ignoring negatives, and keeping away from those with whom students cannot settle disputes. When the problem affects the entire class, it might be helpful to discuss the behavior (not necessarily the problem if it is personal), and seek constructive ways to cope with it. Students often can provide helpful suggestions. Try using minilessons and role playing as methods of introducing social skills. *Note:* If you feel that a dispute or problem between students might lead to violence, inform the appropriate administrator or supervisor.

LIST 284

(Continued)

Problem: The student attempts to dominate groups of which he is a member.

Possible solutions: Place this student in groups with equally strong personalities. Avoid putting him in a group that he can dominate. Speak to this student about appropriate behavior in a group setting, and sit in on groups of which he is a part. Not only will you be able to monitor his behavior closely, but you will be able to model the proper behavior as well.

Problem: The student gives up easily.

Possible solutions: Provide plenty of support and encouragement. Work with this student individually to make sure that she understands how to complete assignments. Also make certain that she has acquired the necessary prerequisite skills to complete the assignments. Offer praise for each small step forward. For group work, place this student with others who will provide her with support.

Problem: You believe that the student is copying homework.

Possible solutions: Speak to the student individually about completing his own work. You may wish to address the entire class on this without singling anyone out. Emphasize that only by doing their own work can students master the skills necessary to complete your course satisfactorily. For students who continue copying, you may need to contact their parents. Remember to be diplomatic here because you are on uneven ground. You may be convinced that a student is cheating, but his mother and father may be equally convinced that he would never do such a thing. Be prepared to answer questions. Noting how his homework is always done perfectly but he has trouble passing a quiz on the same material is a good way to buttress your point. Showing examples of perfectly completed homework and dismal quizzes will help your case. You can then discuss with the parents how important it is for their child to do his own work. Not only will he learn more, but he will gain satisfaction knowing that he is responsible for his good grades.

Problem: The student's work is suffering, and you suspect an undiagnosed learning disability.

Possible solutions: Contact the student's guidance counselor, your school's child study team, or the administrator in charge of learning disabilities and recommend that the student be tested to discover if any learning problem exists. Consult with her previous and other current teachers to find out if they see some of the same problems you do. This can provide you with more anecdotal insight about the problem.

LIST 285

How to Manage a Cooperative Math Class

Cooperative classrooms are based on the belief that students can work together and help each other in the process of learning. The following points will help you to organize your math classes to take advantage of a cooperative atmosphere.

▶ Cooperative learning is based on teams.

- Pairs and groups of three or four work best in the elementary grades.

- In middle school and high school, groups of three to five work well.

- Mix abilities in your groups. Avoid having four top students in one group and four weak students in another.

- Try to balance personalities. Form groups where students are likely to work well together. Do not put an overly dominant student with three shy, quiet ones.

- Mix ethnic groups, and try to balance the number of boys and girls in the groups.

- Change groups periodically.

▶ Explain the purpose of the groups fully. Unless students understand the structure, they may have trouble meeting your expectations.

▶ Assign roles to students, telling them that each student has an important part in the group. Roles you might assign include:

- Leader, who keeps the group on task

- Recorder, who writes down the team's ideas, conclusions, and results

- Time monitor, who keeps track of time

- Materials monitor, who is responsible for any materials the group might need

- Checker, who reviews the group's work

▶ Arrange the classroom furniture to accommodate group work. You might use tables or simply have students slide desks together to form tables.

▶ At the beginning of the year, practice group work. Organize the groups, start them working on a task, and sit in on each group and model the appropriate behavior. Some students may have little experience working in groups and may not know how to work together efficiently and effectively. By assuming the various roles yourself, you will be showing the students how to act.

▶ Set a time limit for group work based on the activity. Remember that work usually expands to fit the amount of time given to do it.

▶ Use a signal, such as switching the lights on and off, ringing a small bell, or clapping your hands, to gain your students' attention when you need to talk to the class or they become noisy.

▶ Allow a few weeks for students to become skilled at working in groups.

Student Guidelines for Working in Groups

Many math activities and problems are ideal for group work. The following suggestions can help any group work more effectively.

- ▶ Group members should be willing to cooperate.
- ▶ Each member should share his or her ideas.
- ▶ Members should refrain from attempting to dominate any discussion.
- ▶ Before speaking, members should carefully think about the points they wish to make. When they speak, they should try to state their ideas clearly.
- ▶ After speaking, individuals should give the floor to others.
- ▶ Members should remember that courtesy and politeness are essential to the smooth working of any group.
- ▶ The discussion should remain focused on math, particularly on the problem to be solved.
- ▶ Listeners should not interrupt speakers. They should note questions and ask them after the speaker is finished.
- ▶ When disagreements arise, they should be discussed calmly, without undue emotion. Everyone should be afforded the opportunity to speak and contribute.
- ▶ Comments should always be kept on the topic and should always be constructive.
- ▶ If members are given specific roles to fulfill (for example, group leader, time monitor, or recorder), each should accept the responsibilities of that role.

LIST 287

Steps for Conducting Effective Conferences with Students

Conferences with students regarding math often lead to increased motivation and achievement on the parts of students. A conference need not be lengthy or formal and may last only a few minutes. It may take place at your desk, at the student's desk, during a free period, or after school.

The purpose of any conference is to help students improve their understanding of mathematics. During the conference, which either you or the student may initiate, you may offer praise for good work, provide encouragement, or explain a specific skill. No matter where or when they occur, conferences will be more successful if you follow the guidelines below.

▶ Try to meet with all of your students periodically. You may be able to speak with only two or three from each class each day, but eventually that will give you the chance to meet with everyone regularly.

▶ Begin the conference by seeking the student's input. Ask him if he is having trouble with anything. If you know he is having problems with a particular skill, use that as your starting point.

▶ Keep the conference focused. Try to address only one or two skills or one type of problem. Trying to do too much can overwhelm and frustrate students.

▶ Build an atmosphere of support and cooperation during the conference. Keep the tone upbeat.

▶ Tailor the conference to meet the individual needs of the student.

▶ During the conference, be ready to assume one of several roles, such as motivator, listener, giver of information, or guide.

▶ Be sincere with your praise. Students quickly realize when praise is false.

▶ Offer positive comments, and always avoid negative or sarcastic remarks. The conference should be a time of support and help.

Copyright © 2005 by Judith A. Muschla and Gary Robert Muschla

LIST 288

Steps to Help Students Develop Problem-Solving Skills

There are many ways you can help students learn and use problem-solving skills. Such skills will help them not only in math class but in the rest of life as well.

▶ Give students realistic, authentic problems that have meaning in their everyday lives.

▶ Give both numerical and nonnumerical problems.

▶ Encourage students to make up their own problems and share them with classmates.

▶ Organize problem-solving groups of three to five students. Distribute problems to the groups, and let students work together and discuss possible methods for solution.

▶ Provide problems that have more than one answer.

▶ Present problems that have unnecessary or missing information. (Have students supply the missing data.)

▶ Give problems that can be solved using mental math.

▶ Encourage students to keep notes of their efforts to solve difficult problems.

▶ Provide problems that require estimation.

▶ Provide an assortment of high-level problems that require explanations in their solutions.

▶ Encourage the use of various problem-solving strategies. (See List 273, "Problem-Solving Strategies for Middle School Students" and List 274, "Problem-Solving Strategies for the Upper Grades.")

▶ Suggest to students that they try to identify various ways to solve a problem and then choose the best one.

▶ Urge students to ask themselves questions as they solve problems.

▶ Consider beginning each class with a problem-of-the-day.

▶ Encourage students to review their progress in solving problems. They should adjust their plan as necessary.

▶ Stress to students that they should check their answers to be sure the answers are logical and make sense.

LIST 289

A Teacher's Plan for Writing Word Problems

While textbooks provide solid word problems for students to solve, it is still likely that you can improve them, or at least tailor them for your class. Here are some suggestions:

- ► Change the original question. Many word problems provide enough information so that you can easily rework the question to fit your needs.

- ► Add extra information to the problem. This forces students to find the necessary information and ignore what is not needed.

- ► Delete some information, and have students identify what other data they need to solve the problem.

- ► Add some information, and create multistep problems out of one-step problems.

- ► Provide students with data from word problems, and ask them to create problems of their own.

- ► Offer hints to help students solve complicated or difficult problems.

- ► Use graphs, charts, and tables from textbooks for data from which students can write problems for each other.

- ► Provide students with the answers to a few problems on a page of problems, and have them find the problems that match the answers given.

- ► Mix problems that have different operations. This requires students to think carefully about which operation is needed.

- ► Ask students to write down the steps they use to solve problems.

Sources of Problems-of-the-Day

Many teachers find it useful to begin class with a problem-of-the-day. Unfortunately, it is sometimes a problem to come up with a problem-of-the-day that corresponds to what you are teaching. Some sources as well as a few tips for creating your own problems-of-the-day follow.

Sources

▶ Newspapers, both local and national. You can find good information that can be turned into relevant problems.

▶ Magazines. Consult news magazines, major monthlies, and the magazines your students read.

▶ Almanacs. These provide plenty of interesting facts and statistics.

▶ The Internet. News features, weather facts, and Web sites on countless topics offer a wealth of information. Math Web sites for teachers are prime sources of ideas. (See List 305, "Math Web Sites for Teachers.")

▶ Your math book. Many math books contain sections of "Extensions" or "Challenges" that you can use for problems-of-the-day.

▶ Books of math starters and do-nows. There are many, one of which is our *Math Starters! 5- to 10-Minute Activities That Make Kids Think, Grades 6–12* (Jossey-Bass, 1999).

▶ Books of math games and puzzles. If your school library does not have these, your local library will.

▶ Major events in school. As students scramble to come up with the cash for prom gowns and tuxedos, you might come up with problems that zero in on money.

Tips for Creating Your Own Problems-of-the-Day

▶ Focus on a review skill. One of the best uses of a problem-of-the-day is to keep students sharp with skills they have already learned.

▶ Make the problems relevant. Tie the problems to a current event of interest to your students or an issue or topic that affects them.

▶ Create problems that have more than one step.

LIST **291**

The Use of Portfolios in Your Classes

A portfolio contains samples of a student's work that is collected over a given length of time. A good portfolio offers insights into a student's thinking, understanding, and mathematical problem-solving skills, and thus provides a picture of the student's overall progress in math. The following suggestions can help you incorporate the use of portfolios in your classroom.

▶ Explain to students what a portfolio is and how it will be used.

▶ Provide students with portfolio envelopes. These should be large enough to hold various kinds of work.

▶ Make sure that your students understand that they are to select their best work for their portfolios. Because one of the purposes of the portfolios is to show individual growth, all papers should be dated.

▶ Portfolios may be one of two kinds: an *assessment* portfolio that shows particular growth, or a *work* portfolio that contains various projects and activities. Generally, papers from the work portfolio are selected to go into the assessment portfolio.

▶ Although you may guide students in their selection of material for their portfolios, they should be the judges of what actually goes in.

▶ The material that goes into a portfolio should help the teacher and others to understand the student's progress in math.

▶ Many papers, activities, and projects are appropriate for inclusion in a portfolio:

- An introduction describing the contents of the portfolio
- A table of contents
- Solutions to difficult problems that detail problem-solving abilities
- Responses to challenging questions and problems
- Applications of mathematics in other subjects
- Integration of math in the student's life
- Problems that the student has created
- An example of the student's group activity
- A written report on a major topic in math
- The student's written account of his or her growth in mathematics

LIST **292**

Helping Students Prepare for Math Tests

Whether it is a unit assessment, an end-of-the-year exam, or a standardized test, there is much you can do to help students prepare.

▶ Be sure that students are familiar with the test's format. They should know how the test is organized, how many sections it will contain, and if there will be a time limit.

▶ Students should be familiar with the types of problems they can expect. It is not fair for students to be given new or trick problems on the test.

▶ Students should have plenty of practice with the types of problems they will be expected to solve on the test.

▶ Lead up to the test by beginning several classes with problems-of-the-day that are similar to the problems that will be on the test. An ongoing review will sharpen students' skills.

▶ If the test requires that students fill in responses on an answer sheet, they should have practice tests that have answer sheets.

▶ Review the day before the test. Encourage students to come to this class with any questions they might have about the test.

▶ Encourage students to get a good night's sleep the night before the test and eat a solid breakfast on the morning of the test.

For additional information on test-taking strategies, see List 271, "Strategies for Taking Math Tests."

LIST 293

Math Bulletin Board Ideas

There is nothing like a good bulletin board to stimulate students' thinking about math. Ideally, your bulletin board should be located near your classroom or in a central location. While every teacher likes to have great bulletin boards, you are probably thinking you do not have the time to come up with good ideas. The following suggestions can help.

▶ Names and photos of top math students. "Math Students of the Month" makes a great display. (Before posting any photographs of your students, be sure to check your school's policy regarding student photographs.)

▶ News about math. Offer information about math contests, math club meetings, conferences, or scholarships, or post articles clipped from newspapers and magazines. You might also post the URLs of interesting math Web sites.

▶ Challenging problems or puzzles. Perhaps offer a "Puzzle of the Week." Be sure to provide the answer for the previous puzzle.

▶ The work of students.

▶ Career information about math. You might post descriptions of jobs that require a strong background and understanding of math and the qualifications needed for those jobs.

▶ Posters showing how mathematics is important to other fields and disciplines.

▶ Examples of mathematical ideas and principles.

▶ Brief biographical sketches about famous mathematicians.

▶ Computer-generated illustrations of mathematical concepts or ideas.

▶ A display of student essays about math; for example, "Math and Me."

▶ A list of colleges that are recognized for their outstanding math departments.

▶ Illustrations or pictures highlighting various geometrical shapes.

LIST 294

Topics for Possible Math Projects

Following are suggestions for math projects.

▶ Create a booklet of original word problems. Produce the booklet on a computer, and distribute copies to the class. (Most software programs have the capability of producing attractive booklets.)

▶ Create a packet of math problems derived from statistics, charts, and tables you find in newspapers and magazines. Distribute copies to the class.

▶ Write a report or give an oral report on a topic in math. Select your own, or consider one of these:

- The History of Mathematics
- Measurement in the Ancient World
- Evolution of the Calendar
- The History of the Magic Square
- Computers and Their Mathematical Applications
- A Famous Mathematician
- Math in Everyday Life
- The Development of Numbers
- The Increasing Role of Mathematics in a Technological World
- Mathematics and Architecture
- Mathematical Principles in Nature
- Math and Art
- Math and Medicine
- Topology
- Vectors
- Matrices
- Inventions and the Use of Mathematics
- Symmetry Patterns in Nature
- Fractals
- Geometry and the City

▶ Choose your favorite sport. Consider how math plays a part in its record keeping and statistics. Present an oral report to the class on the topic.

▶ Write a how-to manual for younger students that explains how to use a calculator effectively.

▶ Create a collage of people using mathematics in their professions.

LIST 294

(Continued)

▶ Design and build models showing geometrical shapes.

▶ Create a scale drawing of the floor plan of your home or, if you are really ambitious, your school.

▶ Create charts and graphs detailing how the school day is broken down into periods and activities.

▶ Create a packet of cross-number puzzles, photocopy it, and distribute it to the class.

▶ Design, create, and produce a newspaper or magazine about mathematics.

▶ Collect magazine articles and newspaper clippings about math and make a portfolio.

▶ Organize a panel discussion on an interesting topic in math. Two good examples are "The Relevance of Math Today" and "The Use of Computers in Modeling Mathematic Concepts."

How to Start a Math Magazine

Math magazines are an excellent way of sharing ideas and news about math. They may be rather simple productions—articles written with basic word processing software, photocopied, and distributed—or the eye-catching result of the latest in desktop publishing, displayed on the school's Web site, and made available for downloading. By far the best are written and produced by students. Several tips for producing quality math magazines follow.

► Although a math magazine can be a class effort, it is usually better to have the magazine produced by the math club. This permits more students from various classes and grades to participate.

► Students should assume most of the responsibility for producing the magazine. Students should act as editors, reporters, researchers, and proofreaders, as well as manage the overall production process.

► You might consider working together with an English teacher. He or she can help students with some of the finer points of writing while you focus on the mathematics.

► Your magazine should come out regularly. Decide whether it will be monthly or quarterly. More often than monthly will be a tough pace to keep up; less than quarterly will make it hard to maintain continuity.

► With your students, decide on what types of articles the magazine will contain. You might focus on math topics, news items regarding mathematics, contests, puzzles, reports of student projects, and interesting problems.

► Start collecting material for the magazine well in advance of deadlines. Set a deadline that will give you enough time for production.

► Encourage students to write their articles on computers using one of the major word processing programs, such as Microsoft Word. They should submit their articles both in hard copy (paper) and on a disk. This will provide much flexibility in revising and editing.

► Illustrate your magazine using the "drawing" components of your software or clipart, which may be a part of your software or which you can easily access on-line.

► Before printing your magazine, edit the articles, and encourage student proofreaders to check every page carefully. Now is the time to catch any remaining mistakes.

► Print enough copies of the magazine for students, faculty, administrators, and displays throughout your school.

► Instead of printing copies, you may choose to publish the magazine on your school's Web site for viewing. Publishing the magazine on a Web site enables people from outside your school to view the magazine. You may also send the magazine via an e-mail attachment to those who wish to have a copy.

LIST 296

How to Start a Math Club

A math club provides an excellent forum for students and teachers to share ideas about mathematics, and listen to outside speakers or members reporting on special topics in math. If you are considering starting a math club in your school, keep the following points in mind.

► The typical math club is sponsored by a faculty member.

► The club's members may consist of students and faculty. Usually students from various classes and grades may join.

► Many math clubs operate under a constitution that details the club's organization and bylaws.

► Meetings may be held whenever the members agree are appropriate times. However, most clubs meet either twice a month or monthly. Less than once a month makes it difficult to maintain the club's purpose; meeting more than twice a month may be a burden for some members.

► Meetings generally run a half-hour to an hour.

► Math clubs often select rather exotic names for themselves—for example, the Euclideans, Pi Squares, or The Circle.

► Many math clubs maintain math libraries to which club members have access.

► Many math clubs undertake various projects, for example:

- Maintaining a mathematics bulletin board
- Publishing a mathematics newspaper or newsletter
- Sponsoring math contests in school
- Providing tutoring programs for students who request help in math
- Constructing models, diagrams, and posters highlighting topics in mathematics and distributing them throughout the school
- Providing a forum for recognizing student achievement in math
- Helping with a parents' math night

Ideas for Math Field Trips

Field trips offer a way to show students how math is used in the real world. Following is a list of possible field trips for your math classes.

Museum—Look for exhibits that show how math is used in our lives.

Science center—See how math is used in science.

Amusement park—Note the principles of math as they relate to the attractions.

Supermarket—See the ways numbers are used in pricing and calculating.

Accounting firm—Learn how accountants work with numbers.

Post office—Learn how postal rates are calculated and how math relates to packaging and transport.

Engineering firm—Learn how formulas are used in design.

Bank—Learn how interest rates and percentages are used in banking and finance.

Architectural firm—See how architects use formulas in the design of buildings.

Nursery (gardening)—Learn how math is used to design flower beds and how math is used in purchasing and selling plants.

Insurance office—Find out how insurance rates are determined.

Carpentry shop—Learn how measurements are used in construction.

Toy manufacturer—See how math is used in the design and manufacture of toys.

Real estate office—Learn how commissions are determined as a percentage of the selling price of a home.

Newspaper—See how math is used in the design of the copy.

Lumber yard—Find out how divisibility rules are used in cutting materials to produce the least amount of waste.

Sign shop—See how geometry is used in designing signs.

Print shop—See how geometry is used in making posters and displays.

Retail store—Learn how math is used in determining sale prices and discounts.

LIST 298
Possible Guest Speakers for Your Math Classes

Some students have trouble recognizing the relevance of math. They do not see how important math is in their lives. Bringing guest speakers into your classes can often help students realize that an understanding of mathematics is essential. The list below suggests some people you might ask to speak to your classes about the importance of math.

Actuary—Explain how formulas are used to determine insurance rates.

Auto mechanic—Discuss the use of metric measurements in car repairs or explain how math is used in measuring liquid capacities for fuel, oil, power steering fluid, brake fluid, and other liquids.

Bank officer—Speak about savings, loans, and interest rates.

Stockbroker—Explain how securities are bought and sold.

Architect—Explain how geometry is used to design buildings.

Computer programmer—Speak about the mathematical foundations of computers.

Advertising agent—Discuss how statistics are used in determining target audiences.

Builder/contractor—Speak about how formulas are used in constructing buildings.

Engineer—Discuss how formulas are used in determining the weight limit of bridges.

Astronomer—Explain scientific notation and light-years for representing distances in astronomy.

Accountant—Discuss sales, income, and other taxes.

Newspaper editor—Explain how statistics are gathered and used by newspapers in articles.

Chef—Talk about how fractions are used in cooking and baking.

Pharmacist—Explain how metric measurements are used in labeling the strength or amount of medications.

Biologist—Explain how sampling can be used to determine animal or plant populations in a given area.

Interior decorator—Explain how measurement plays a vital role in furnishing a room or home.

Statistician—Explain how various types of data are gathered and analyzed.

Photographer—Discuss shutter speeds on cameras.

Retail salesperson—Explain how percentages and discounts are determined.

Marketing specialist—Discuss how consumer markets are evaluated.

Pollster—Discuss the design and use of random samples.

LIST **299**

Steps for Maintaining Positive Relations with Parents

Enlisting the support of parents throughout the year can go a long way in helping students achieve their potential in your classes. The suggestions below can aid you in your efforts to enlist the support of the parents of your students.

▶ View parents as positive resources. Although it is true that some parents are their children's biggest "problem," the great majority are truly interested in their children's education and want to help.

▶ Be available to parents. Return telephone calls and answer notes promptly.

▶ If your school has a Web site and teachers have e-mail accounts, use e-mail as a way parents can contact you and you can contact them. *Note:* For most teachers, it is not advisable to make personal e-mail addresses available to parents or students.

▶ At back-to-school night, along with informing parents about your math program, also inform them that you appreciate any support they can give their children in learning math. Explain how they can contact you.

▶ Throughout the year, keep parents informed of their children's progress. Obviously, report cards and interim notices do this, but there are other ways as well. When a student does exceptionally well on a test, drop a line home. If a student begins to slip, do not wait for the mid-marking period notices: contact his or her parent. The use of e-mail can be a big help here. Most parents want to know about such changes in their children's performance.

▶ If your school maintains a Web site where teachers can post assignments, study tips, or announcements, be sure you use the Web site and keep information current.

▶ During conference time, be prepared to meet with parents. Have available any notes, copies of tests, quizzes, portfolios, and homework assignments that support what you need to say.

▶ Be specific in discussing the weaknesses and strengths of your students and how parents can help their children.

▶ Be tactful when speaking with parents. Saying that Johnny "never does a lick of homework and is as sloppy as his old man" is unlikely to result in a positive or useful conference.

▶ Avoid sugarcoating. Do not tell a parent that if her daughter works exceptionally hard, she has a chance for an A when it is unlikely she will be able to earn more than a C. Building false hopes only sets up people for a letdown.

▶ Notify parents of upcoming standardized tests.

▶ Explain how parents can help their children prepare.

▶ Share with parents your enthusiasm for mathematics, and educate them about the importance of learning math.

LIST **300**

How Parents Can Help
Their Children Learn Math

As their children advance through the grades, many parents begin to feel that they can no longer help them with their math work. Either the math is becoming too complicated, they do not want to show their children the way they were taught, which might be different from the methods their children's teachers use, or they feel that by helping their children learn math, they will be assuming the teacher's role. Whatever the reason, when parents take a back seat, their children lose a valuable resource. Some simple things all parents can do to help children learn math are listed below.

▶ Become acquainted with the math teacher at back-to-school night and parent-teacher conferences.

▶ Let the math teacher know that you want to help your child at home. Ask what specific help you might provide. Most teachers are glad to offer suggestions.

▶ Encourage your child to complete his or her math neatly and accurately. This helps to reduce careless mistakes.

▶ Encourage your child to restate word problems, especially what information is given and what the question is asking.

▶ Practice estimating with your child. When you go to the store to buy a few items, ask your child to estimate the cost.

▶ Practice measurement. Estimate the time necessary for trips, and estimate the heights and weights of various objects. When you are measuring things around the house, such as new curtains for windows, encourage your child to help.

▶ Never tell your child that some people "just aren't good at math," or that you aren't "good" with math. Instead, emphasize that competency can be gained through hard work.

▶ Never think that girls cannot do math as well as boys.

▶ Always reward your child with praise when he or she does well. That helps to build confidence.

▶ Talk about math in a positive manner. Point out the many practical tasks that require an understanding of mathematics, such as planning a home budget, calculating grocery bills, or setting enough money aside for a vacation.

▶ Whenever possible, use math at home in problem solving. Encourage your child to work out the problem.

▶ Encourage your child to use logic and common sense when solving math problems.

LIST 301

Back-to-School-Night Checklist

The first contact you will have with many of the parents of your students is at back-to-school night. Since first impressions are often difficult to change, you should strive to make the session positive for both you and your visitors. The following list can help you make back-to-school night successful.

▶ Preparation is essential. Prior to the big night, plan and rehearse your presentation. In the typical back-to-school night, the teacher introduces himself or herself and provides a brief summary of the classroom program and policies, including:

- Attendance

- Goals

- Expectations

- Homework

- Grading

- Assessment

- How technology will be used in the classroom

- Important dates, such as those for standardized testing

▶ Make sure your classroom is clean and orderly. Display the work of your students, and have available examples of the textbooks, workbooks, and technology students will be using.

▶ On back-to-school night, dress professionally and arrive early. Greet parents at the door to the classroom, and shake their hands. Ask them to sign a sign-in sheet, which should be placed at the door. It is a good idea to have two columns: one for the parents' names and the next for the name of their son or daughter.

▶ During the presentation, be positive and enthusiastic. Do not mention specific students or any problems you might be experiencing with the class.

▶ Be prepared for parents who will ask about their own child. Leave your grade book at home, and explain that you are not prepared to discuss individual students at this time. Offer to arrange a conference at a later date to discuss any personal concerns they have. Note that back-to-school night is a time for a general presentation.

▶ Tailor your presentation to fit the allotted time. Trying to pack in too much will leave parents with the impression that you are rushed and disorganized; providing too little may leave them with the idea that you will not be using all of the class time to teach their children. Some teachers prefer to leave some time for parents to ask questions. *Be warned:* Opening the session to questions can result in parents' asking about things you are not prepared to discuss. If this happens, suggest that you set up a separate meeting to discuss the issue. A thorough presentation should answer most questions.

LIST 301

(Continued)

▶ Consider using technology to enhance your presentation. Displaying classroom rules on an overhead projector or using PowerPoint to highlight your talk can add professionalism.

▶ Prepare a packet of handouts. Parents appreciate being able to take home printed materials summarizing your program, policies, and contact information.

▶ Ask for the support of your students' parents. Encourage them to contact you with any questions, and explain how they can reach you at school by telephone or e-mail. (It is not advisable to provide your personal telephone number or e-mail address.) You should check for messages on a regular basis.

LIST 302

Steps for Conducting Effective Parent-Teacher Conferences

Parent-teacher conferences provide an opportunity for you to share the progress of a student with his or her parents. It is a time when you can encourage parents to help their children succeed in your math class. There are several steps you can take to ensure the effectiveness of your conferences with parents.

▶ Prior to the conference, review the student's grades and work. Make sure you have a goal for the outcome of the conference. Perhaps the student needs to be more diligent with completing homework, study more for tests, or keep focused on task.

▶ Write down the points you would like to cover. Note the student's strengths and weaknesses that you would like to cover. Some teachers have a folder for each student in which they keep notes and samples of the student's work.

▶ Collect samples of the student's work to show the parent.

▶ Anticipate some of the questions the parent might ask you. Consider answers. These actions will help reduce the chances of your being surprised and taken off guard.

▶ Arrange chairs around a table. Avoid sitting behind your desk, which can give parents a feeling that you are aloof.

▶ Have paper and pens handy to write down notes, both for yourself and the parent.

▶ Start the conference by thanking the parent for coming, and try to establish a friendly, comfortable atmosphere. Remember that both you and the parent share the same purpose of helping the child succeed in your math class. Offer some positive remarks about the student.

▶ State any problems or concerns you have simply, offering facts to support your statements. Show examples of work that illustrate any problems, or give details of inappropriate behavior. Be sure to explain the actions you have taken to solve the problem.

▶ During the conference, always remain calm and professional.

▶ If parents become defensive or upset, remind them that both of you must work together in order to help their child.

▶ Establish a plan to resolve the problem the student is facing. Clearly state what you will do and what the parent should do.

▶ At the end of the conference, review the points that were discussed, and tell the parent how he or she can contact you. Explain how you will follow up the conference—perhaps a telephone call in two weeks—to discuss the student's progress. Remember to thank the parent for coming to the conference.

▶ After the conference, be sure to write down any pertinent notes, and be sure to follow up.

A note of caution: Do not talk about other students or parents during the conference, and do not share any confidential information.

LIST 303

How to Have a Parents' Math Night

Parents want their children to enjoy school and be successful. Most realize the importance of mathematics and would like to help their children with their work, but they are not sure how. A math night for parents can show them how to support your efforts in the classroom, as well as emphasize the many ways mathematics affects our lives.

Plan in Advance

► Clear the idea of having a math night for parents with your principal or supervisor. Enlist his or her support.

► Decide who your audience will be; for example, parents of elementary students or parents of middle or high school students. You might aim for parents of a specific grade level. If you do not limit your audience, you risk not being able to present relevant activities. The needs and program of a fourth-grade math student are quite different from those of a tenth grader.

► Decide who will be invited. Parents of just your school? The entire district? Will you include private and parochial schools of your town?

► Decide how many spaces will be available. If you will be working alone, an audience of more than twenty-five will be hard to manage, especially if your activities include manipulatives. You will be able to handle more parents if you have help, perhaps a colleague or student volunteers. Student council or math club members are usually good choices.

► Choose a date. Check your school calendar to make sure that there are no conflicts with sports events, concerts, or other programs. Generally, nights around major holidays, community activities, or vacations are not good times. Perhaps tie parents' math night to a general PTA meeting, which will add to your turnout.

► Obtain funds for supplies, promotions, and refreshments. Know how much you can spend, and do not exceed your budget.

► Publicize your parents' math night in district, school, and PTA newsletters and other publications. List the date on the school calendar, announce it at any meetings you attend, and provide press releases to local newspapers. You should also send flyers home with students. If you are limiting the number of participants, be sure to inform parents that they must sign up in advance. Providing a confirmation slip on the flyer is an easy way to get this information.

► Decide what activities you will provide. Ideally, activities should help parents to understand the math their children are learning, various teaching techniques, problems students might be encountering, and ways parents can support their children's efforts in math. Manipulatives provide hands-on experience while also illustrating concepts.

LIST 303

(Continued)

▶ Decide what equipment you will use. Will you use an overhead projector in your presentation? Will you need a computer? Will you need computers for parents? Make certain you are familiar with any software you will be using. Arrange for the equipment to be available and functioning properly for math night.

▶ Create any packets of information you will be handing out to parents.

The Big Night

▶ Arrive early, and make sure that you are organized. Set up the refreshments at the back of the room. If a colleague or students will be helping you, meet with them, and review what they are to do.

▶ Check any equipment you will be using. Make sure that the overhead projector is plugged in and that the computer you will be using for your PowerPoint presentation is booted up. Pull the projection screen down.

▶ Make any handouts available to parents as they enter. This eliminates the need for taking time to pass out materials during your presentation.

▶ To begin the session, introduce yourself and state your objectives for the evening. If the group is small enough, ask parents to introduce themselves.

▶ Briefly explain your math curriculum and how the activities you planned for this night will support the learning efforts of the parents' children.

▶ Move into the first part of your session as quickly and smoothly as possible. A parents' math night should be activity oriented.

▶ Plan a break at about the halfway point. Encourage parents to help themselves to refreshments. During the break, make yourself available to parents. If many seem to have the same questions, address them in the second part of your presentation. Limit the break to around ten minutes.

▶ Provide time for parents' questions at the conclusion of the session. Hand out an evaluation form that asks parents how your math night might be improved.

▶ Provide follow-up as necessary. You will likely find the parents' math night to be as enlightening to parents as it is rewarding to you.

LIST 304

Professional Math Organizations

Although many organizations provide excellent information and opportunities to those interested in mathematics, the following are likely to be of most interest to teachers and math supervisors.

American Mathematical Society

201 Charles Street

Providence, RI 02940-2294

800-321-4AMS (United States and Canada)

401-445-4000 (worldwide)

Fax: 401-331-3842

www.ams.org

Association for Women in Mathematics

4114 Computer and Space Sciences Building

University of Maryland

College Park, MD 20742-2461

301-405-7892

Fax: 301-314-9074

www.awm-math.org

Mathematical Association of America

1529 18th Street, NW

Washington, DC 20036-1385

800-741-9415

202-387-5200

Fax: 202-265-2384

www.maa.org

National Council of Supervisors of Mathematics

P.O. Box 150368

Lakewood, CO 80215-0368

Voice mail and fax: 303-274-5932

http://mathforum.org/ncsm

LIST **304**

(Continued)

National Council of Teachers of Mathematics

1906 Association Drive

Reston, VA 20191-1502

703-620-9840

Fax: 703-476-2970

www.nctm.org

School Science and Mathematics Association

www.ssma.org

LIST 305

Math Web Sites for Teachers

There are many Web sites for math teachers. A search using "Web sites for math teachers" on any of the major Internet search engines will likely yield enough hits to keep you busy for quite some time. To narrow the search, we have included in the following list some of what we believe are the best Web sites for math teachers. Note that many of these sites have links to other excellent sites.

▶ *National Council of Teachers of Mathematics:* http://www.nctm.org

▶ *FunBrain.com:* www.funbrain.com/teachers/index.html

▶ *Mega Mathematics:* www.c3.lanl.gov/mega-math/

▶ *The Math Forum:* http://mathforum.org

▶ *Interactive Mathematics, Miscellany, and Puzzles:* www.cut-the-knot.org/shtml

▶ *Algebra.help:* www.algebrahelp.com/content.jsp

▶ *About Mathematics:* http://math.about.com

▶ *Platonic Realms:* www.mathacademy.com/pr/index.asp

▶ *The Prime Pages:* http://primes.utm.edu

▶ *Eisenhower National Clearinghouse:* www.enc.org

▶ *PBS Mathline:* www.pbs.org/teachersource/math.htm

▶ *Math.com:* www.math.com

▶ *Go Math:* http://gomath.com

▶ *S.O.S. Mathematics:* www.sosmath.com/

▶ *Education World:* www.educationworld.com/math

▶ *Figure This!:* www.figurethis.org/teacher_corner.htm

LIST **306**

A Self-Appraisal for Math Teachers

Answering the questions in this list can help you improve your math classes and make them an interesting, challenging, and satisfying experience for your students.

- ► Have I established clear and consistent policies and procedures that support my instructional objectives?

- ► Do I communicate to my students that through hard work, everyone—regardless of gender or background—can do well at math?

- ► Do I maintain a positive atmosphere in my classes?

- ► While being demanding and challenging in my requirements, am I also considerate of the individual needs of my students?

- ► Do I provide my students with the instruction and activities that will help them to master the skills and concepts in my curriculum?

- ► Do I provide interesting and engaging activities?

- ► Does my instruction satisfy the Standards of the NCTM?

- ► Do I encourage the use of technology in my classes?

- ► Do I provide problems whose solutions require analysis and critical-thinking skills?

- ► Do I provide time and opportunities for both individual and group activities?

- ► Are my methods of evaluation fair and consistent?

- ► Do I encourage and respond to students' questions?

- ► Do I maintain positive relationships with colleagues, supervisors, and parents?

- ► Do I encourage parents to support their children in the learning of math?

- ► Am I aware of and responsive to the various needs of my students?

- ► Do I always treat math as an important subject?

Reproducible Teaching Aids

The reproducibles that follow are designed to support various programs and needs and can be adapted to fit your lessons and teaching style. For example, "Number Lines" saves you the trouble of making number lines. Some of the samples on the sheet have been labeled, while others are not, allowing you to put in your own values. "Decimal Squares" shows your students the relationship between wholes, tenths, and hundredths. "Blank Checks" and "Blank Check Register" provide you with premade checks and a register you can use to demonstrate to your students the proper way to write and record checks.

The other reproducibles can be just as helpful. In some cases, you may wish to create transparencies to use with an overhead projector or make your own manipulatives by cutting out the designs and laminating them.

Number Lines

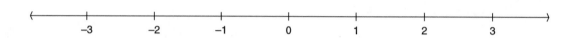

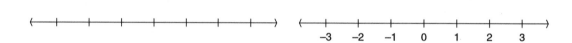

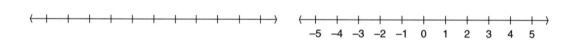

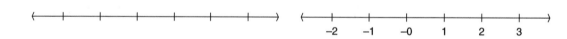

Rulers I

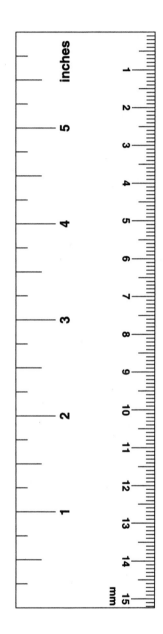

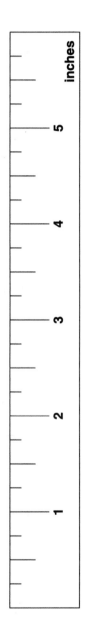

Rulers II

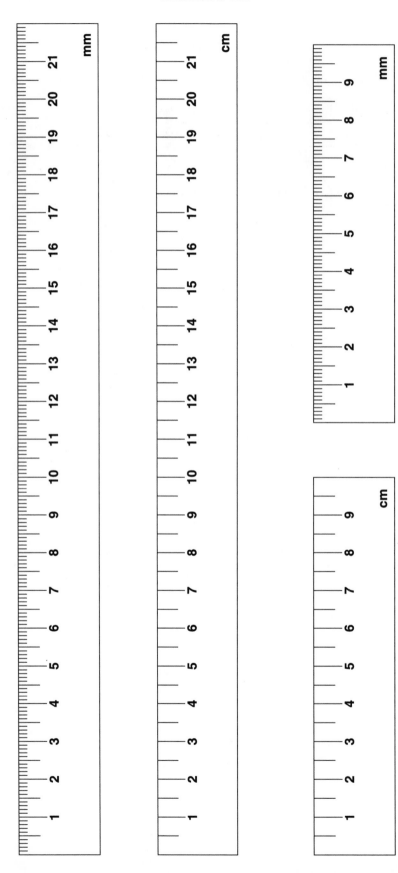

Protractors

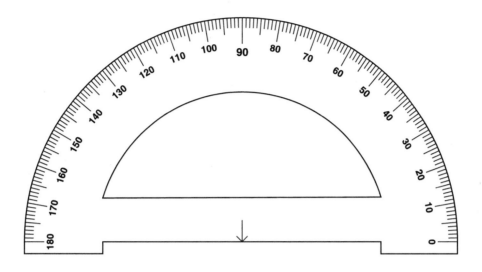

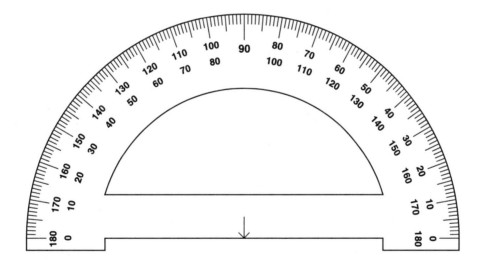

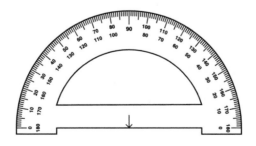

Squares

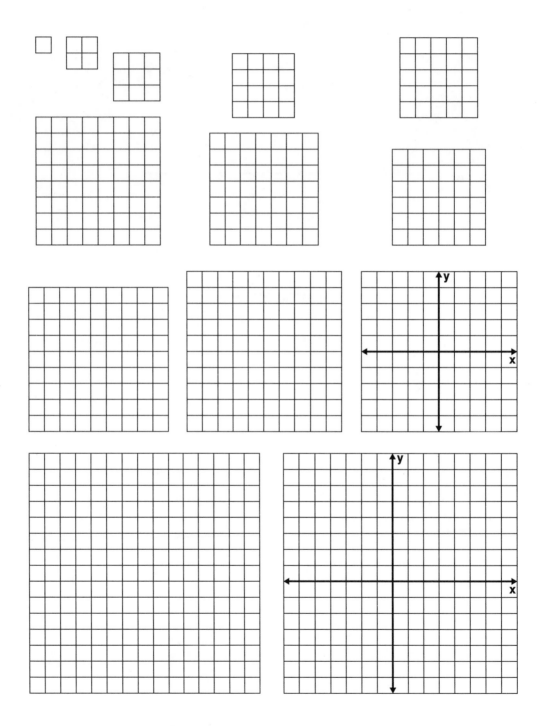

Graphs for Trigonometric Functions

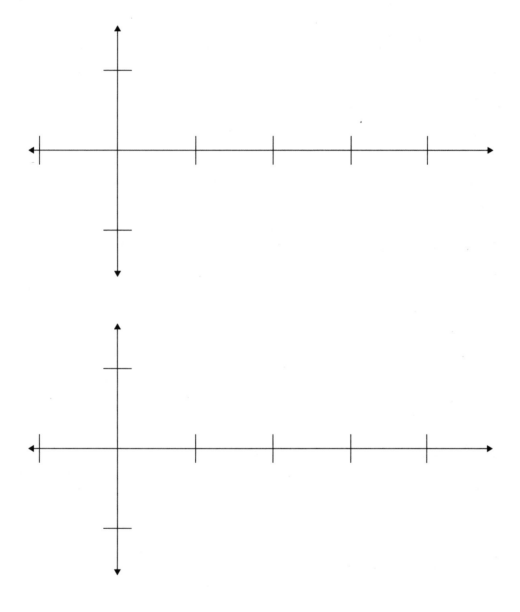

1-Inch Grid

$\frac{1}{2}$-Inch Grid

1-Cm Grid

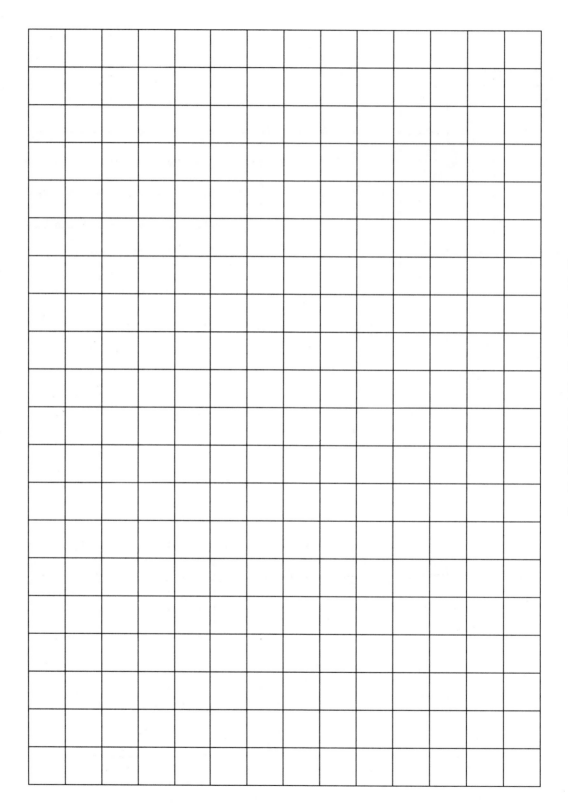

Isometric Dot Paper

Square Grid Dot Paper

Decimal Squares

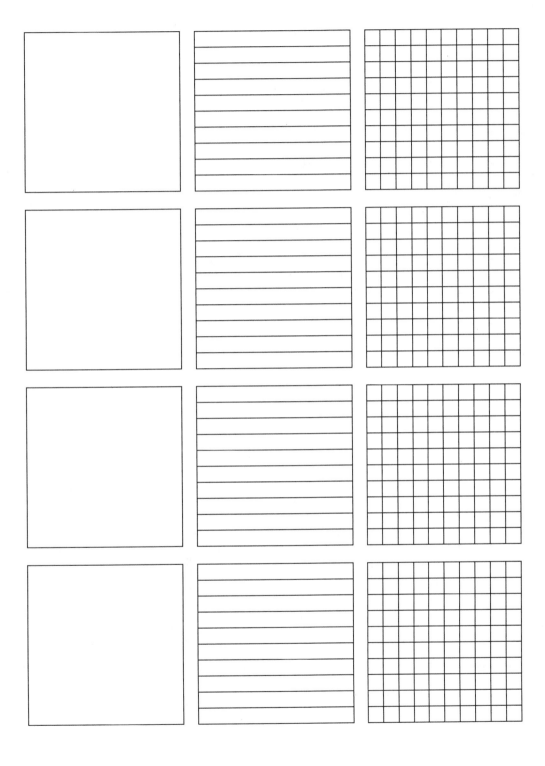

Fraction Strips

1											

| $\frac{1}{2}$ | | | | | | $\frac{1}{2}$ | | | | | |

| $\frac{1}{3}$ | | | | $\frac{1}{3}$ | | | | $\frac{1}{3}$ | | | |

| $\frac{1}{4}$ | | | $\frac{1}{4}$ | | | $\frac{1}{4}$ | | | $\frac{1}{4}$ | | |

| $\frac{1}{5}$ | | $\frac{1}{5}$ | | $\frac{1}{5}$ | | $\frac{1}{5}$ | | $\frac{1}{5}$ | | | |

| $\frac{1}{6}$ | | $\frac{1}{6}$ | | $\frac{1}{6}$ | | $\frac{1}{6}$ | | $\frac{1}{6}$ | | $\frac{1}{6}$ | |

| $\frac{1}{8}$ | $\frac{1}{8}$ | $\frac{1}{8}$ | $\frac{1}{8}$ | $\frac{1}{8}$ | $\frac{1}{8}$ | $\frac{1}{8}$ | $\frac{1}{8}$ | | | | |

| $\frac{1}{9}$ | $\frac{1}{9}$ | $\frac{1}{9}$ | $\frac{1}{9}$ | $\frac{1}{9}$ | $\frac{1}{9}$ | $\frac{1}{9}$ | $\frac{1}{9}$ | $\frac{1}{9}$ | | | |

| $\frac{1}{10}$ | $\frac{1}{10}$ | $\frac{1}{10}$ | $\frac{1}{10}$ | $\frac{1}{10}$ | $\frac{1}{10}$ | $\frac{1}{10}$ | $\frac{1}{10}$ | $\frac{1}{10}$ | $\frac{1}{10}$ | | |

| $\frac{1}{12}$ | $\frac{1}{12}$ | $\frac{1}{12}$ | $\frac{1}{12}$ | $\frac{1}{12}$ | $\frac{1}{12}$ | $\frac{1}{12}$ | $\frac{1}{12}$ | $\frac{1}{12}$ | $\frac{1}{12}$ | $\frac{1}{12}$ | $\frac{1}{12}$ |

Fraction Circles

Algebra Tiles

Tangram

Patterns for Pattern Blocks

Patterns for Cuisenaire® Rods

Net for Rectangular Prism

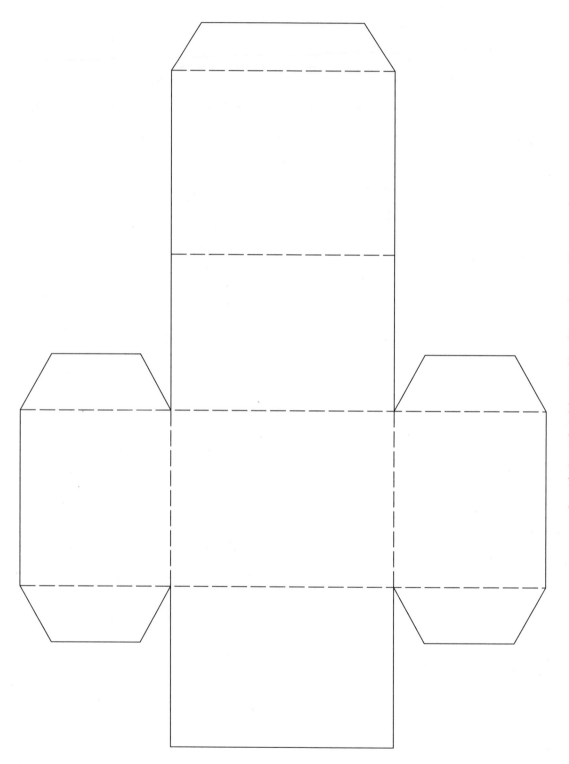

Net for Pyramid

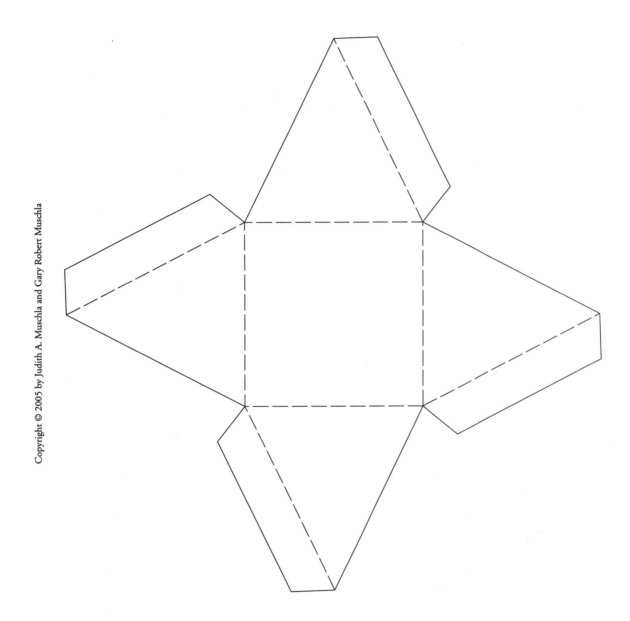

Net for Cone

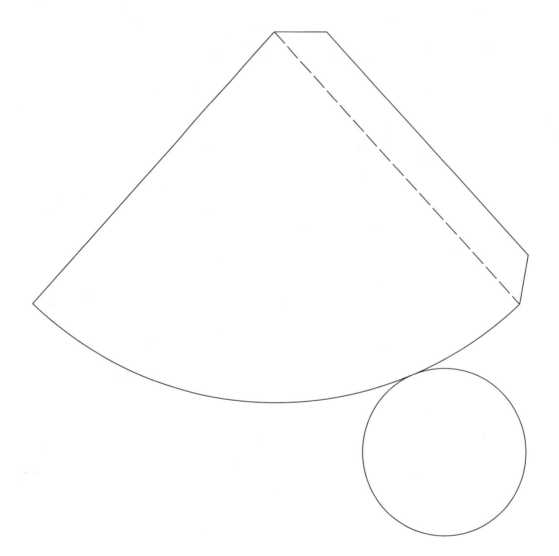

Net for Tetrahedron

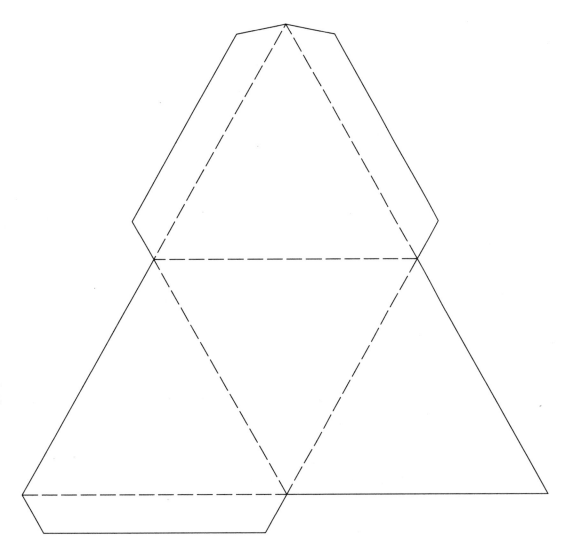

Net for Hexahedron (Cube)

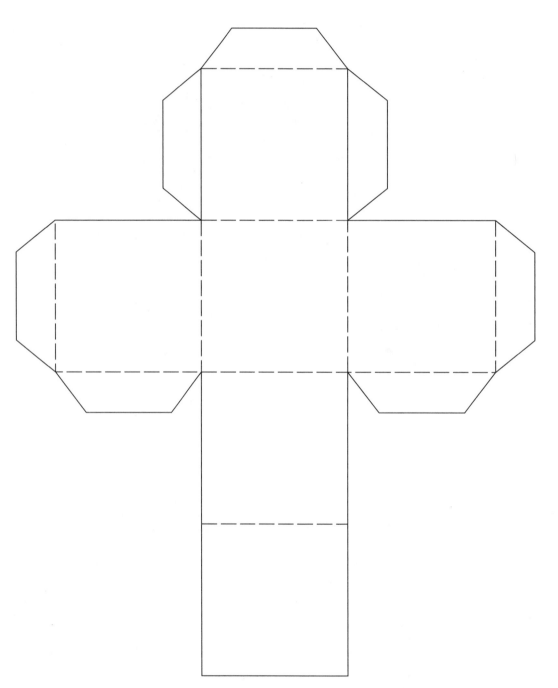

Net for Octahedron

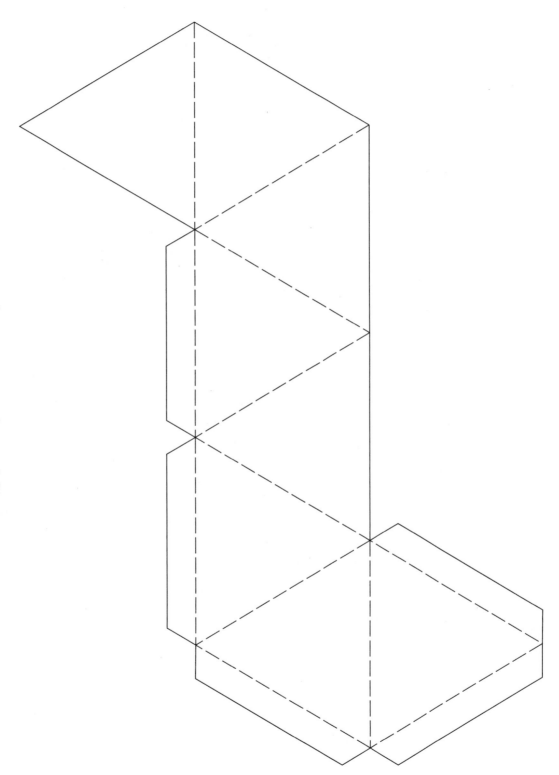

Net for Dodecahedron

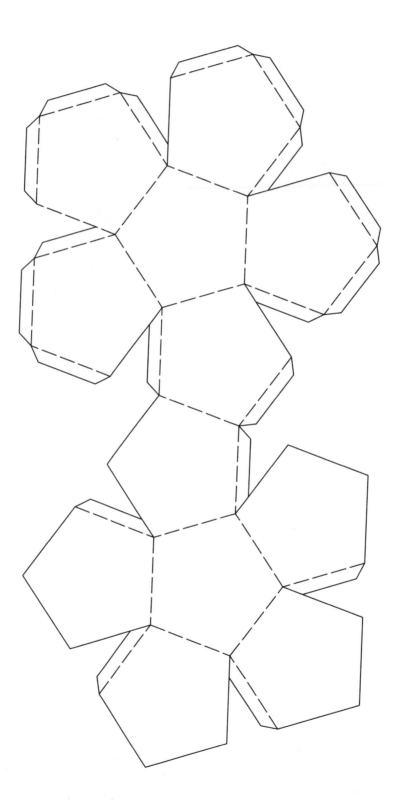

Net for Icosahedron

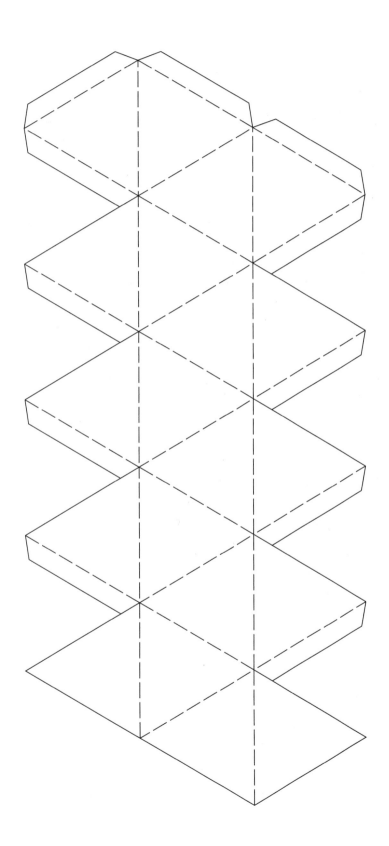

Blank Checks

John E. and Jane Doe	101

_____ 20 ____

PAY
TO THE
ORDER OF _____ $ _____

_____ DOLLARS

Strongbox Savings
South River Office
South River, NJ 08882

FOR _____   _____

John E. and Jane Doe	102

_____ 20 ____

PAY
TO THE
ORDER OF _____ $ _____

_____ DOLLARS

Strongbox Savings
South River Office
South River, NJ 08882

FOR _____   _____

John E. and Jane Doe	103

_____ 20 ____

PAY
TO THE
ORDER OF _____ $ _____

_____ DOLLARS

Strongbox Savings
South River Office
South River, NJ 08882

FOR _____   _____

Blank Check Register

						Balance
Check Number	Date	Description of Transaction	Payment Debit	✔	Deposit/ Credit	

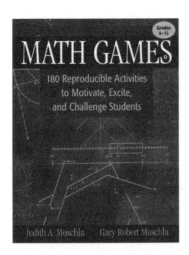

Math Games:
180 Reproducible Activities to Motivate, Excite, and Challenge Students, Grades 6-12

Judith A. Muschla and Gary Robert Muschla

Paper   ISBN: 0-7879-7081-6
www.josseybass.com

This is a dynamic collection of 180 reproducible activity sheets to stimulate and challenge your students in all areas of math from whole numbers to data analysis while emphasizing problem solving, critical thinking, and the use of technology for today's curriculum.

Each of the book's activities can help you teach students in grades 6 through 12 how to think with numbers, recognize relationships, and make connections between mathematical concepts. You pick the activity appropriate for their needs, encourage the use of a calculator, or provide further challenges with activities that have multiple answers.

Designed to be user friendly, all of the ready-to-use activities are organized into seven convenient sections and printed in a lay-flat format for ease of photocopying as many times as needed.

- Whole Numbers - Fractions, Decimals, and Percents - Geometry - Measurement
- Algebra - Data Analysis - Potpourri

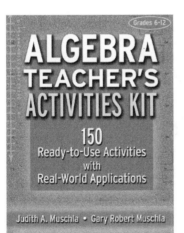

Algebra Teacher's Activities Kit:
150 Ready-to-Use Activities with Real-World Applications

Judith A. Muschla and Gary Robert Muschla

Paper   ISBN: 0-7879-6598-7
www.josseybass.com

A unique resource that provides 150 ready-to-use algebra activities designed to help students in grades 6-12 master pre-algebra, Algebra I, and Algebra II. The book covers the skills typically included in an algebra curriculum. Developed to motivate and challenge students, many of the activities focus on real-life applications. Each of the book's ten sections contains teaching suggestions that provide teachers with strategies for implementing activities and are accompanied by helpful answer keys. The activities supply students with quick feedback, and many of the answers are self-correcting.

Each activity stands alone and can be applied in the manner that best fits your particular teaching program. *Algebra Teacher's Activities Kit* can be used as a supplement to your instructional program, to reinforce skills and concepts you've previously taught, for extra credit assignments, or to assist substitute teachers.

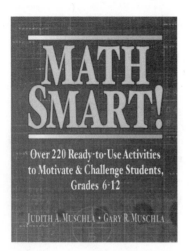

Math Smart!
Over 220 Ready-to-Use Activities to Motivate & Challenge Students, Grades 6-12

Judith A. Muschla and Gary Robert Muschla

Paper   ISBN: 0-7879-6642-8
www.josseybass.com

Here's a super collection of 222 reproducible activity sheets to stimulate and challenge your students in all areas of math from whole numbers to data analysis while emphasizing problem solving, critical thinking, and the use of technology for today's curriculum!

These activities teach students how to think with numbers, recognize relationships, and make connections between mathematical concepts. You pick the activity appropriate for their needs ... Let them self-check their answers ... encourage the use of a calculator ... or provide further challenges with activities that have multiple answers. For quick and easy use, all of these ready-to-use activities are organized into seven convenient sections:

- Whole Numbers
- Fractions, Decimals & Percents
- Measurement
- Geometry
- Algebra
- Data Analysis
- Potpourri

Hands-On Math Projects With Real-Life Applications: Ready-to-Use Lessons and Materials for Grades 6-12

Judith A. Muschla and Gary Robert Muschla

Paper   ISBN: 0-13-032015-3
www.josseybass.com

Help students apply math concepts and skills to everyday problems found across the curriculum, in sports, and in daily life with this outstanding collection of 60 hands-on investigations. These tested projects stress cooperative learning, group sharing, and writing as they build skills in problem solving, critical thinking, decision-making and computation. What's more, you get tested guidelines, techniques and tools for managing the classroom during project activities and assessing students' performance.